Forensic Taphonomy and Ecology of North American Scavengers

Forensic Taphonomy and Ecology of North American Scavengers

Susan N. Sincerbox
Texas State University, San Marcos, TX, United States

Elizabeth A. DiGangi
Binghamton University, Binghamton, NY, United States

ACADEMIC PRESS

An imprint of Elsevier

Academic Press is an imprint of Elsevier
125 London Wall, London EC2Y 5AS, United Kingdom
525 B Street, Suite 1800, San Diego, CA 92101-4495, United States
50 Hampshire Street, 5th Floor, Cambridge, MA 02139, United States
The Boulevard, Langford Lane, Kidlington, Oxford OX5 1GB, United Kingdom

British Library Cataloguing-in-Publication Data
A catalogue record for this book is available from the British Library

Library of Congress Cataloging-in-Publication Data
A catalog record for this book is available from the Library of Congress

ISBN: 978-0-12-813243-2

For Information on all Academic Press publications
visit our website at https://www.elsevier.com/books-and-journals

Working together
to grow libraries in
developing countries

www.elsevier.com • www.bookaid.org

Publisher: Mica Haley
Acquisition Editor: Elizabeth Brown
Editorial Project Manager: Sam Young
Production Project Manager: Priya Kumaraguruparan
Cover Designer: Christian Bilbow

Typeset by MPS Limited, Chennai, India

CONTENTS

LIST OF FIGURES

LIST OF TABLES

ACKNOWLEDGMENTS

We are indebted to Dr. Sèbastien Lacombe for his valiant work on the distribution maps, and we acknowledge Anne Hull for her contribution. We are equally indebted to David Tuttle, who cheerfully snapped photographs and provided several of his own, frequently after a last-minute request. We thank R.T. Kramer, Jessica C. Galea, and Lauren A. Meckel; as well as Drs. Daniel Wescott and Natalie Langley for providing photographs. We are also grateful to Krystle N. Lewis for allowing us to photograph remains being used for her thesis research, specifically those depicted in Figs. 5.15 and 5.16. We thank Drs. Kathleen Sterling, Bill Belcher, Andy Merriwether, and Peter Knuepfer for providing helpful comments; Drs. Jonathan Bethard and Megan Moore for being useful sources of bibliographic information and advice; and Dr. Nicole Siegel for simplifying information about quadruped anatomy and providing information about jaguar scavenging. Dr. Lubna Omar provided access to several animal skeletal specimens for photography and clarified taxonomic information for some of them. Several Binghamton University students provided editorial assistance, such as searching for references or photographs or general fact-finding. We are grateful to Helen Brandt, Sierra Cotrona, Sarah Cunningham, Ariel Gruenthal-Rankin, Grace Kilroy, Danielle Julien, Samantha Kombel, and Kimberly Raymond for taking time out of their busy schedules to apply their considerable Internet search skills. We also thank Anne Larrivee who clarified citation rules for open access images. Finally, we thank our editors at Elsevier, Liz Brown, Joslyn Chaiprasert-Paguio, Sam Young, and Project Manager Priya Kumaraguruparan, for agreeing to take on this project and their patience as we worked to complete it.

Susan N. Sincerbox thanks her undergraduate mentor and coauthor, Dr. Elizabeth A. DiGangi, for inspiring her pursuit of a career in forensic anthropology, for encouraging and facilitating the publication of this book, and for always going above and beyond for her students, colleagues, and friends. She also acknowledges Dr. Anne B. Clark for inspiring the interdisciplinary nature of this project. She would also like to thank Dr. Roderick Davis for providing invaluable insight into the ecological concepts presented throughout this text. Susan would also like to thank her family and friends, especially her parents for their unwavering support through 23 years of personal milestones and professional endeavors; her best friend and significant other, Liam Davis, for being her biggest champion as she juggled her never-ending to-do list; and Lucy, for curtailing her Frisbee regimen to spend afternoons inside, providing much-needed comic relief while Susan wrote.

Elizabeth A. DiGangi acknowledges her education at the University of Tennessee, and especially the teachings of Dr. Walter Klippel, both having inspired her interest in forensic taphonomy. She thanks Simón and Luna for obliging to leave typical canid marks on the nonhuman bones in Figs. 5.4, 5.8, and 5.9; for their patience on some of the nicest days of the summer while she worked on this manuscript instead of taking them to the park; and as always, for their companionship.

CHAPTER 1

Introduction

FORENSIC SCIENCE UNDER THE MEDICOLEGAL MICROSCOPE
FORENSIC ERROR: PROVING GUILT IN THE INNOCENT
PUTTING "SCIENCE" IN FORENSIC SCIENCE
CONCLUSION
GUIDE TO THIS VOLUME
REFERENCES

Anthropology is the study of human beings. Given this broad scope, the field is divided into four major subdisciplines to fully answer the question of what it means to be human: (1) archaeology, (2) biological anthropology, (3) cultural anthropology, and (4) linguistic anthropology. Forensic anthropology is a relatively young subfield within biological anthropology which originated when anthropological knowledge of human skeletal biology began to be applied to medicolegal questions. By the 1970s, around the time that the discipline was formalized via establishment of professional organizations in the United States, such questions were limited to decedent identification and were addressed using osteological markers and measurements. Since the field's inception, however, the role of the forensic anthropologist has expanded beyond identification to include evidence recovery and documentation, skeletal trauma analysis, and interpretation of postmortem modifications to skeletal material (Dirkmaat et al., 2008).

Thus, contemporary forensic anthropologists are concerned not only with the identification of the decedent, but with reconstruction of postmortem events including deposition of remains and decomposition. Forensic taphonomy emerged as a subfield as forensic anthropologists increasingly recognized the importance of context to analyzing human skeletal remains. Taphonomy means "laws of burial" and was first developed by paleontologists to help them understand what processes

Forensic Taphonomy and Ecology of North American Scavengers. DOI: https://doi.org/10.1016/B978-0-12-813243-2.00001-4

and variables would lead to fossilization, important when attempting to locate possible fossil deposits (Efremov, 1940). Paleoanthropologists began applying taphonomic principles to their own studies to answer questions about early hominid behavioral patterns, such as whether populations were engaged in hunting or scavenging of large prey, evidenced by the types of bone modifications present (i.e., marks left by carnivore scavenging vs marks left by stone tools) (Binford, 1981; Blumenschine, 1986, 1988; Brain, 1983). With the expansion of forensic anthropology's purview discussed earlier, taphonomy was incorporated to better understand the archaeological context of a burial or surface scatter and the factors that may have impacted the remains since deposition, including decomposition.

Early research in taphonomy attempted to identify sequential patterns in human decomposition that could be used to estimate the postmortem interval, or time since death, through retrospective studies of forensic case reports (Galloway et al., 1989; Megyesi et al., 2005). Researchers quickly learned that decomposition is incredibly complex, with the rate and patterning impacted by both intrinsic variables (i.e., body size, presence of disease or trauma) and environmental variables (i.e., climate, animal scavenging) (Mann et al., 1990). To address these challenges, decomposition research facilities, including the University of Tennessee's Anthropology Research Facility in Knoxville, Tennessee and the Forensic Anthropology Center at Texas State University in San Marcos, Texas, were created to conduct actualistic studies of human decomposition in outdoor contexts.

Scavenging by vertebrate taxa, or animals, is common when human remains have been deposited outdoors, and can also occur indoors, when pets or pests are present. Significant research has been conducted in an attempt to describe the taphonomic signatures (i.e., marks left behind) of different species, including: vultures (Beck et al., 2015; Reeves, 2009; Spradley et al., 2012), bears (Carson et al., 2000; Murad and Boddy, 1987; Saladié et al., 2013), dogs and coyotes (Colard et al., 2015; Haglund, 1997a, Haglund et al., 1989; Steadman and Worne, 2007; Willey and Snyder, 1989), wild cats (Montalvo et al., 2007; Pickering and Carlson, 2004), rodents (Haglund, 1992; Haglund, 1997b; Klippel and Synstelien, 2007; Tsokos et al., 1999), and other small mammals (King et al., 2016; Synstelien, 2013; Synstelien and Klippel, 2005). Most taphonomy textbooks include chapters summarizing these signatures for major scavenging taxa such as canids and rodents (Haglund and Sorg, 1997, 2001; Pokines and Symes, 2014; Schotsmans et al., 2017). However, for many taxa that are less frequently involved in forensic casework, scavenging behavior and taphonomic effects are often extrapolated from a handful of case reports or incidental observations made in larger scavenging studies.

Additionally, the data produced in forensic scavenging studies are inherently context-specific due to behavioral nuances between populations of the same species across space and time (i.e., differences in vulture (*Coragyps atratus* and *Cathartes aura*) scavenging patterns between Central Texas and Southern Illinois (Dabbs and Martin, 2013); differences in the food habits of American puma (*Felis concolor*) populations in temperate compared to subtropical habitats (Iriarte et al., 1990); and different activity patterns and dietary shifts by coyotes (*Canis latrans*) in response to the reintroduction of wolves (*Canis lupus*) into northern Montana (Arjo and Pletscher, 2000)). Intraspecies variation in scavenging patterns has seen scant treatment in the forensic literature, as it has been traditionally assumed that all members of a scavenging species will follow predictable patterns of behavior. The above examples demonstrate that this assumption may not always hold true. For instance, in their forensically-focused study of vulture behavior, Dabbs and Martin (2013) found that differences in behavior were reflected in the taphonomic signature left by the birds on human remains. Therefore, understanding the behavioral variability of scavengers is critical in accurately interpreting and identifying taphonomic signatures.

An ecological approach to vertebrate scavenging behavior would allow inferences to be made about scavenged death scenes in regions where scavenger behavior has yet to be forensically studied. While existing forensic research provides a foundation for case analysis and interpretation, an understanding of specific ecological influences on scavenger behavior is important to shape effective, individualized strategies for recovery and analysis of human remains. The ecological principles presented in this book, coupled with detailed profiles of "typical" behavior of scavenging organisms, may be drawn upon to provide a theoretical underpinning for decisions made in the course of a death investigation. Sound theoretical bases for investigative decision-making and forensic interpretation contribute to successful casework by bolstering the reliability of forensic analyses and, as will be discussed in the following section, support decisions of admissibility of evidence in a courtroom setting.

FORENSIC SCIENCE UNDER THE MEDICOLEGAL MICROSCOPE

Evidentiary standards play a significant role in shaping the research agenda in the forensic sciences. Consequently, no text on the subject is complete without a brief discussion of the historical evolution of evidence admissibility criteria. In the United States, the first criterion for determining the admissibility of scientific evidence was the Frye standard, also referred to as the Frye rule or Frye test. James Alphonso Frye had been convicted of second-degree murder, and appealed the conviction claiming that expert testimony concerning a passed polygraph test was inappropriately deemed inadmissible. In *Frye v. United States*, 293 F. 1013 (D.C. Cir. 1923), the Court of Appeals for the District of Columbia determined that the conclusions presented as expert testimony must be based on scientific principles and methods which are sufficiently established to be admissible in court. In short, the Frye standard requires that the scientific basis of forensic evidence be generally accepted within the field. The Court determined that polygraph tests did not yet satisfy this requirement and thus were inadmissible, affirming the conviction.

In *Daubert v. Merrell Dow Pharmaceuticals, Inc.*, 509 US 579 (1993), the US Supreme Court clarified that Federal Rule of Evidence (FRE) 702 superseded the Frye standard. The Federal Rules of Evidence were drafted by the US Supreme Court and passed by the US Congress in 1975 to address evidence introduction at federal trial. FRE 702 specifically establishes legal requirements for admissibility of expert witness testimony. In a California District Court, two families sued Merrell Dow pharmaceutical company, claiming that a drug produced to treat morning sickness had caused serious birth defects in their sons. The defense and prosecution both presented expert witnesses with conflicting testimony concerning the drug's status as a potential teratogen. The District Court would not admit the scientific evidence presented by the prosecution, determining that the in vitro and in vivo animal studies were not yet generally accepted and, therefore, did not meet the criteria of the Frye standard.

Upon appeal, the US Supreme Court ruled that FRE 702 had superseded the Frye standard. FRE 702 requires that expert witness testimony aid the jury in determining facts at issue, that the testimony is based on sufficient data analyzed with reliable methods that are interpreted in the light of reliable principles, and that this analysis and interpretation are applied to the facts of the case (Rule 702). These requirements can be summarized as two key criteria: (1) relevance, and (2) reliability. Ultimately, the *Daubert* case established these criteria for the admissibility of scientific evidence and positioned the trial judge as responsible for evaluating adherence to these criteria through review of the scientific principles and methodologies employed by both the prosecution and the defense. Following this decision, the criteria demanded by FRE 702 are often informally referred to as the "*Daubert* standard."

Within a decade of the *Daubert* decision, the Supreme Court decided two additional cases concerning the admissibility of scientific evidence: *General Electric Company v. Joiner*, 522 US 136 (1997) and *Kumho Tire Company v. Carmichael*, 526 US 137 (1999). In *General Electric Company v. Joiner*, an electrician developed small-cell lung cancer following years of work-related exposure to polychlorinated biphenyls (PCBs) and proceeded to sue his employer. Experts were called upon by both the defense and prosecution, presenting conflicting testimony concerning whether a causal link could be established between PCB exposure and small-cell cancer development in humans. The District Court ruled that the prosecution's expert testimony was inadmissible because it failed to demonstrate a link between cancer and exposure to PCBs. However, this decision was reversed by an appellate court, ruling that the District Court had abused their discretion by excluding this evidence on the basis of the expert's conclusions alone. Ultimately, the Supreme Court ruled that the District Court had not abused its discretion in denying the admission of the prosecution's evidence. This final ruling reaffirmed the trial judge's role as gatekeeper of evidence admissibility, to include discretion with regard to exclusion of expert testimony solely on the basis of the conclusions.

In *Kumho Tire Company v. Carmichael*, the Carmichael family brought suit after a tire on the family's vehicle blew out, killing one passenger and resulting in significant injury to others. The family called in an expert in tire failure analysis, whose testimony would support their assertion that the tire was defective. Applying the *Daubert* standard, the District Court found that the expert's testimony was unreliable because it could not be replicated, had not been subject to peer review, and had no potential error rate. When the Carmichael family appealed, the appellate court reversed the decision on the grounds that the testimony did not depend on scientific principles and thus should not be subject to the *Daubert* standard. However, the Supreme Court ruling favored a flexible interpretation of the *Daubert* decision, allowing application of FRE 702 to expert testimony reliant on other types of specialized knowledge.

Expert witnesses are required to evaluate forms of specialized evidence that may otherwise be difficult for those outside technical fields to understand. The *Daubert* trilogy—namely, *Daubert*, *Joiner*, and *Kumho* decisions together—therefore serves as a form of quality control for expert witness testimony, promoting the admissibility of evidence derived from valid scientific or technical principles and methods. Ultimately, the *Daubert* trilogy standard aims to prevent the presentation of unfounded conclusions from evidence analysis in court, ideally reducing the likelihood of a misguided court decision.

FORENSIC ERROR: PROVING GUILT IN THE INNOCENT

Errors arising from flawed practices in the forensic sciences can have life or death consequences. The legal system of the United States assumes an individual's innocence, leaving the burden on the prosecution to demonstrate an individual is guilty beyond a reasonable doubt. The nature and magnitude of the relationship between forensic error and criminal case processing has proved difficult to evaluate, but nevertheless has been the subject of numerous studies. In a study of over 4000 cases, Peterson et al. (2013) found that forensic evidence had a robust influence on the processing of criminal casework, finding that arrest rate was significantly influenced by evidence collection, while conviction rate was significantly influenced by subsequent laboratory analysis.

However, Baskin and Sommers (2013) criticized the Peterson et al. (2013) study for aggregating all cases statistically without regard to the nature of the crime or the forensic evidence collected, noting that crimes of different natures (e.g., burglary vs homicide) are subject to different types of evidence collection and methods of investigation and therefore cannot be statistically compared. In a previous study using the same data, Baskin and Sommers (2010) evaluated variables influencing case processing for homicides committed by adult offenders. Of 400 homicide cases included in the study, physical evidence was collected in 97%, analyzed in 81%, and autopsies were performed in approximately 72%. However, despite the presence of forensic evidence, only 34.5% of cases were resolved with a conviction. Baskin and Sommers (2010) suggested that conviction rates may instead, increase when the relationship between perpetrator and victim is close, or when eyewitnesses are available. In the United States, this is consistent with a demonstrated decrease in conviction rates that has occurred in tandem with an increase in homicides committed by strangers (Baskin and Sommers, 2010; Wellford and Cronin, 2000).

The jury is still out on exactly how forensic evidence collection and interpretation impacts the probability of rendering a verdict of innocence vs. guilt. However, flawed forensic science has been demonstrated to be a contributor to erroneous outcomes (Garrett and Neufeld, 2009; Gould et al., 2014). Suspects who are guilty may be acquitted and released, or innocent suspects may be convicted. While little information is available on forensic error and improper acquittals, substantial research has examined the role that forensic error has played in wrongful convictions. In a study of 200 individuals exonerated by postconviction DNA analysis, Garrett and Neufeld (2009) found that in 57% of the cases, forensic evidence supporting a guilty conviction was presented. For these cases, serological analysis of blood or semen was the most common form of evidence presented, and this forensic evidence was a major component of the prosecution's case. In a study of criminal casework in the United States, Gould et al. (2014) identified forensic error as one of several significant predictors of wrongful conviction.

Both Garrett and Neufeld (2009) and Gould et al. (2014) noted that in cases resulting in wrongful convictions, flawed forensic science served to compound other sources of error such as mistaken eyewitness testimony or coerced confessions. These correlations demonstrate a form of "tunnel vision" in investigators and other medicolegal officials, i.e., skewed interpretation so that some forms of evidence are ignored and other forms are made to fit with existing evidence (Gould et al.,

2014). In forensic analyses, this may present as confirmation bias (Kassin et al., 2013). Eerland et al. (2012) suggests that the interpretation of forensic evidence is likely influenced by an additional bias known as the feature positive effect, that people have more difficulty processing absent information. Eerland et al. (2012) used an experimental design to evaluate how individuals determined guilt when presented with mixed information regarding present (i.e., bloodstains at the scene) and absent (i.e., the missing firearm) forensic evidence. In justifying their decisions of guilt or innocence, study participants were significantly more likely to draw on present evidence while ignoring absent evidence.

The conviction of innocent suspects can have particularly dire consequences in states where the death penalty is punitive. These consequences are compounded by an increased likelihood of erroneous conviction of individuals indicted in such states (Gould et al., 2014). Gould et al. (2014) attributes this correlation to the legal culture of death penalty states which encourages swift resolution and, consequently, hasty evaluation of evidence. Following accusations of forensic malpractice, human rights organizations such as the Innocence Project have assumed the task of revisiting and reviewing the forensic evidence supporting allegedly false convictions.

PUTTING "SCIENCE" IN FORENSIC SCIENCE

In November 2005, the United States Congress authorized the National Academy of Sciences (NAS) to create a dedicated committee to evaluate current forensic practices and determine their reliability and validity (National Academy of Sciences, 2009). Part of the impetus for this was the fallout from the misidentification of an American citizen's fingerprint on bomb making materials in Madrid, Spain after the 2004 train bombings. The resulting report, *Strengthening Forensic Science: A Path Forward*, produced by the newly established Committee on Identifying the Needs of the Forensic Sciences Community, identified several issues within forensic science to include: the lack of standardization of methods, the diversity and poor compatibility of forensic disciplines, and the failure of common practices to produce reliable and valid results.

All in all, the NAS report (2009) demonstrated that many practices which had become standard in forensic investigation—including friction ridge analysis in fingerprint identification, questioned document examination, and evaluation of bite marks—had not been demonstrated to be reliable or scientifically valid. Consequently, pressure was placed on the different disciplines in forensic science to vastly improve their methods while simultaneously reevaluating their theoretical underpinnings. In the meantime, such traditional forms of evidence became vulnerable to dismissal from court proceedings.

To address some of the NAS report (2009) recommendations, the National Institute of Standards and Technology (NIST), part of the US Department of Commerce, was tasked in 2013 by the US Department of Justice to establish and maintain some form of working groups that would establish best practices in the forensic sciences. This relationship was established due to NIST's overall mission of promoting standards in science and technology. Previously, scientific working groups for different forensic disciplines existed, but all were not under the same organizational umbrella. This new arrangement aims to standardize working group organization and productivity. The resultant configuration consists of multiple levels. The Organization of Scientific Area Committees (OSAC) for Forensic Science consists of five Scientific Area Committees (SAC) that themselves consist of discipline subcommittees, with all levels overseen by the Forensic Science Standards Board. The Anthropology subcommittee is under the SAC for Crime Scene/Death Investigation.

The NAS report (2009) and the OSACs continue to exert pressure on American forensic scientists to reform their methodologies in accordance with new evidentiary standards. In forensic anthropology, standardization of methods has long been on the radar (e.g., Buikstra and Ubelaker, 1994), and initiatives to develop quantitative methods for skeletal identification had already begun in response to changing evidentiary standards or repatriation laws (e.g., Christensen, 2004, 2005; Steadman et al., 2006). Following the NAS report (2009), research seeking to validate and improve traditional identification methodologies through the use of advanced statistical analyses has flourished (including but not limited to: Anderson et al., 2010; Edgar, 2013; Hefner and Ousley, 2014; Kenyhercz et al., 2014; Klales et al., 2012; Moore et al., 2016; Shirley and Ramirez-Montes, 2015; Spradley et al.,

2015; Spradley and Jantz, 2011). Additionally, computer software such as Fordisc (Ousley and Jantz, 2005), 3Skull (Ousley, 2004), CRANID (Wright, 2008), and Osteoware ("Osteoware: Standardized skeletal documentation," 2016) have been developed to promote standardization of skeletal analysis, albeit with some limitations.

Given the complexity of taphonomic scenarios, standard methods for data collection and interpretation in this arena have been slower to develop. To estimate postmortem interval (PMI), Megyesi et al. (2005) attempts to quantify the relationship between ambient temperature and a measure of gross decomposition referred to as the total body score. Although this decomposition scoring system has low interobserver error, the total body score has been demonstrated to be a poor predictor of PMI cross-regionally and cross-seasonally due to differences in the decomposition environment (Dabbs et al., 2016; Marhoff et al., 2016; Suckling et al., 2016; Sutherland et al., 2013). Alternative methods of scoring gross decomposition have additionally been proposed, including the Degree-Day Index (Michaud and Moreau, 2011), Degree of Decomposition Index (Marhoff et al., 2016), and the Accumulated Decomposition Score system (Gleiber et al., 2017). Researchers have also examined a variety of variables contributing to variation in human decomposition to include those from the environment (i.e., rainfall, insect colonization) and those intrinsic to the individuals themselves (i.e., body mass, preexisting trauma, presence of clothing) (Archer, 2004; Bates and Wescott, 2016; Cross and Simmons, 2010; Mann et al., 1990; Matuszewski et al., 2014; Rodriguez and Bass, 1983; Simmons et al., 2010).

Scavenging by animals has long been recognized as having an impact on death scene investigations. However, most species scavenge as only one facet of a more complex foraging strategy, and thus it is a difficult phenomenon to systematically study in natural settings. For many species, forensic knowledge of scavenging behavior has been extrapolated from individual case reports or auspicious observations as part of unrelated studies. Exceptions include canids and rodents, the taphonomic signatures of which have been researched extensively given their frequent involvement in forensic casework (Haglund, 1992, 1997a, b; Haglund et al., 1989; Klippel and Synstelien, 2007; Pokines et al., 2016; Tsokos and Schulz, 1999; Willey and Snyder, 1989). Animal scavenging behavior in other taxa remains understudied with regard to its effects on forensic questions such as PMI estimates and perimortem trauma analysis.

CONCLUSION

Although significant progress has been made in the forensic sciences, error persists. A 2016 report to President Barack Obama revealed uncomfortably high error rates in traditional lines of feature-comparison evidence, including analysis of bite marks, latent fingerprints, firearms, footwear impressions, and hair—many of the same areas that were identified as problematic in the 2009 NAS report (President's Council of Advisors on Science and Technology, 2016). Similar to the NAS report (2009), the President's Council of Advisors on Science and Technology (2016) called for increased research to verify the validity of traditional forensic methods, the development of objective analytical methods, and the development and standardization of nationally recognized best practices. While taphonomy has yet to be specifically addressed in national reports, forensic investigators conducting such analyses should follow suit in ensuring the validity and (reasonable) objectivity of their approaches.

GUIDE TO THIS VOLUME

This book provides forensic investigators with guidelines for dealing with vertebrate-scavenged and scattered remains. It compiles literature from multiple disciplines to form a holistic resource on the taphonomic effects produced by the foraging behavior of the vertebrate scavenging guild in North America, defined as the group of animals that exploit carrion as a foraging strategy. Chapter 2, *Unwitting Accomplices: Scavengers and Forensic Investigation*, serves as an introduction to the impact that scavengers can have on death investigations, and discusses human identification, trauma analysis, and postmortem interval in light of possible scavenger damage. Chapter 3, *There's No Such Thing as a Free Lunch: The Evolution of Scavenging*, covers basic principles of scavenger evolution and ecology, followed by a discussion in Chapter 4, *Scavenger Identification Strategies: Interpreting Taphonomic Signatures*, of how vertebrate morphology, physiology, and behavior impact the taphonomic alterations they produce. Chapter 5, *What Big Teeth You Have: Taphonomic Signatures of North American Scavengers*, presents a field-friendly literature review drawn from numerous academic disciplines including archaeology, forensic anthropology, and behavioral ecology for major scavenging taxa (i.e., biologically-defined groups, such as species) in North America. Chapter 6, *Ecological Influences on Scavenging Behavior*, discusses ecological influences on scavenger behavior including climate, community composition (i.e., local species present) and anthropogenic activity. In Chapter 7, *Adapting Your Investigation: Recovery and Interpretation*, we argue that archaeological techniques are ideal for scene investigation while making suggestions for best practices in shaping strategies to maximize recovery of scavenged remains and improve forensic reconstruction of perimortem and postmortem events. Finally, Chapter 8, *Suggestions for Future Directions*, serves to conclude the volume by suggesting future research directions. In addition, we include an appendix with a glossary of relevant technical and anatomical terminology, as well as a brief overview of the human skeletal system for non-experts and an abridged bibliography of recommended reading.

In light of past and ongoing medicolegal challenges to forensic evidence, this book identifies how practitioners can develop a decision-making framework for investigating the complicated decomposition scenarios presented by animal-scavenged remains. However, given the complexity of ecological relationships that influence scavenger behavior, decisions must ultimately be made on a case-by-case basis and consulting a biologist or ecologist, when possible, is strongly recommended. We hope that these guidelines contribute to improved forensic investigation outcomes by increasing the recovery rate of scattered human remains and associated evidence, improving estimations of the postmortem interval, and reducing error in the forensic reconstructions of perimortem and postmortem events.

REFERENCES

Anderson, M.F., Anderson, D.T., Wescott, D.J., 2010. Estimation of adult skeletal age-at-death using the Sugeno fuzzy integral. Am. J. Phys. Anthropol. 142 (1), 30−41.

Archer, M.S., 2004. Rainfall and temperature effects on the decomposition rate of exposed neonatal remains. Sci. Just. 44 (1), 35−41.

Arjo, W.M., Pletscher, D.H., 2000. Behavioral responses of coyotes to wolf recolonization in northwestern Montana. Can. J. Zool. 77 (12), 1919−1927.

Baskin, D., Sommers, I., 2010. The influence of forensic evidence on the case outcomes of homicide incidents. J. Crim. Justice. 38 (6), 1141−1149.

Baskin, D.R., Sommers, I., 2013. Commentary on: Peterson J.L., Hickman M.J., Strom K.J., Johnson D.J. Effect of forensic evidence on criminal justice case processing. J. Forensic. Sci. 58 (5), 1401–1402.

Bates, L.N., Wescott, D.J., 2016. Comparison of decomposition rates between autopsied and non-autopsied human remains. Forensic. Sci. Int. 261, 93–100.

Beck, J., Ostericher, I., Sollish, G., De Leon, J., 2015. Animal scavenging and scattering and the implications for documenting the deaths of undocumented border crossers in the Sonoran Desert. J. Forensic. Sci. 60 (S1), S11–S20.

Binford, L.R., 1981. Bones: Ancient Men and Modern Myths. Academic Press, San Diego.

Blumenschine, R.J., 1986. Carcass consumption sequences and the archaeological distinction of scavenging and hunting. J. Hum. Evol. 15 (8), 639–659.

Blumenschine, R.J., 1988. An experimental model of the timing of hominid and carnivore influence on archaeological bone assemblages. J. Archaeol. Sci. 15 (5), 483–502.

Brain, C.K., 1983. The Hunters or the Hunted? An Introduction to African Cave Taphonomy. University of Chicago Press, Chicago.

Buikstra, J.E., Ubelaker, D.H. (Eds.), 1994. Standards for Data Collection for Human Skeletal Remains. Arkansas Archaeological Survey Research Series, Fayetteville.

Carson, E.A., Stefan, V.H., Powell, J.F., 2000. Skeletal manifestations of bear scavenging. J. Forensic. Sci. 45 (3), 515–526.

Christensen, A.M., 2004. The impact of Daubert: Implications for testimony and research in forensic anthropology (and the use of frontal sinuses in personal identification). J. Forensic. Sci. 49 (3), 427–430.

Christensen, A.M., 2005. Testing the reliability of frontal sinuses in positive identification. J. Forensic. Sci. 50 (1), 18–22.

Colard, T., Delannoy, Y., Naji, S., Gosset, D., Hartnett, K., Bécart, A., 2015. Specific patterns of canine scavenging in indoor settings. J. Forensic. Sci. 60 (2), 495–500.

Cross, P., Simmons, T., 2010. The influence of penetrative trauma on the rate of decomposition. J. Forensic. Sci. 55 (2), 295–301.

Dabbs, G.R., Martin, D.C., 2013. Geographic variation in the taphonomic effect of vulture scavenging: The case for southern Illinois. J. Forensic. Sci. 58 (S1), S20–S25.

Dabbs, G.R., Connor, M., Byrheway, J.A., 2016. Interobserver reliability of the total body score system for quantifying human decomposition. J. Forensic. Sci. 61 (2), 445–451.

Dirkmaat, D.C., Cabo, L.L., Ousley, S.D., Symes, S.A., 2008. New perspectives in forensic anthropology. Am. J. Phys. Anthropol. 137 (S47), 33–52.

Edgar, H.J.H., 2013. Estimation of ancestry using dental morphological characteristics. J. Forensic. Sci. 58 (S1), S3–S8.

Eerland, A., Post, L.S., Rassin, E., Bouwmeester, S., Zwaan, R.A., 2012. Out of sight, out of mind: The presence of forensic evidence counts more than its absence. Acta. Psychol. (Amst). 140 (1), 96–100.

Efremov, I.A., 1940. Taphonomy: A new branch of paleontology. Pan-American Geologist 74 (2), 81–93.

Galloway, A., Birkby, W.H., Jones, A.M., Henry, T.E., Parks, B.O., 1989. Decay rates of human remains in an arid environment. J. Forensic. Sci. 34 (3), 607–616.

Garrett, B.L., Neufeld, P.J., 2009. Invalid forensic science testimony and wrongful convictions. Va. Law. Rev. 95, 1–97.

Gleiber, D.S., Meckel, L.A., Siegert, C.C., McDaneld, C.P., Pyle, J.A., Wescott, D.J., 2017. Accumulated decomposition score (ADS): An alternative method to total body score (TBS) for quantifying gross morphological change associated with decomposition. Proc. Am. Acad. Foren. Sci. 23, 206–207.

Gould, J.B., Carrano, J., Leo, R.A., Hail-Jares, K., 2014. Predicting erroneous convictions. Iowa. Law. Rev. 99 (2), 471–522.

Haglund, W.D., 1992. Contribution of rodents to postmortem artifacts of bone and soft tissue. J. Forensic. Sci. 37 (6), 1459–1465.

Haglund, W.D., 1997a. Dogs and coyotes: Postmortem involvement with human remains. In: Haglund, W.D., Sorg, M.H. (Eds.), Forensic Taphonomy: The Postmortem Fate of Human Remains. CRC Press, Boca Raton, pp. 367–382.

Haglund, W.D., 1997b. Rodents and human remains. In: Haglund, W.D., Sorg, M.H. (Eds.), Forensic Taphonomy: The Postmortem Fate of Human Remains.. CRC Press, Boca Raton, pp. 405–414.

Haglund, W.D., Sorg, M.H. (Eds.), 1997. Forensic Taphonomy: The Postmortem Fate of Human Remains. CRC Press, Boca Raton.

Haglund, W.D., Sorg, M.H. (Eds.), 2001. Advances in Forensic Taphonomy: Method, Theory, and Archaeological Perspectives. CRC Press, Boca Raton.

Haglund, W.D., Reay, D.T., Swindler, D.R., 1989. Canid scavenging/disarticulation sequence of human remains in the Pacific Northwest. J. Forensic. Sci. 34 (3), 587–606.

Hefner, J.T., Ousley, S.D., 2014. Statistical classification methods for estimating ancestry using morphoscopic traits. J. Forensic. Sci. 59 (4), 883–890.

Iriarte, J.A., Franklin, W.L., Johnson, W.E., Redford, K.H., 1990. Biogeographic variation of food habits and body size of the America Puma. Oecologia. 85 (2), 185–190.

Kassin, S.M., Dror, I.E., Kukucka, J., 2013. The forensic confirmation bias: Problems, perspectives, and proposed solutions. J. Appl. Res. Mem. Cogn. 2 (1), 42–52.

Kenyhercz, M.W., Klales, A.R., Kenyhercz, W.E., 2014. Molar size and shape in the estimation of biological ancestry: A comparison of relative cusp location using geometric morphometrics and interlandmark distances. Am. J. Phys. Anthropol. 153 (2), 269–279.

King, K.A., Lord, W.D., Ketchum, H.R., O'Brien, R.C., 2016. Postmortem scavenging by the Virginia opossum (*Didelphis virginiana*): Impact on taphonomic assemblages and progression. Forensic. Sci. Int. 266, 576.e1–576.e6.

Klales, A.R., Ousley, S.D., Vollner, J.M., 2012. A revised method of sexing the human innominate using Phenice's nonmetric traits and statistical methods. Am. J. Phys. Anthropol. 149 (1), 104–114.

Klippel, W.E., Synstelien, J.A., 2007. Rodents as taphonomic agents: Bone gnawing by brown rats and gray squirrels. J. Forensic. Sci. 52 (4), 765–773.

Mann, R.W., Bass, W.M., Meadows, L., 1990. Time since death and decomposition of the human body: Variables and observations in case and experimental field studies. J. Forensic. Sci. 35 (1), 103–111.

Marhoff, S.J., Fahey, P., Forbes, S.L., Green, H., 2016. Estimating post-mortem interval using accumulated degree-days and a degree of decomposition index in Australia: A validation study. Austr. J. Foren. Sci. 48 (1), 24–36.

Matuszewski, S., Konwerski, S., Frątczak, K., Szafaøowicz, M., 2014. Effect of body mass and clothing on decomposition of pig carcasses. Int. J. Legal. Med. 128 (6), 1039–1048.

Megyesi, M.S., Nawrocki, S.P., Haskell, N.H., 2005. Using accumulated degree-days to estimate the postmortem interval from decomposed human remains. J. Forensic. Sci. 50 (3), 618–626.

Michaud, J.P., Moreau, G., 2011. A statistical approach based on accumulated degree-days to predict decomposition-related processes in forensic studies. J. Forensic. Sci. 56 (1), 229–232.

Montalvo, C.I., Pessino, M.E.M., González, V.H., 2007. Taphonomic analysis of remains of mammals eaten by pumas (*Puma concolor* Carnivora, Felidae) in Central Argentina. J. Archaeol. Sci. 34 (12), 2151–2160.

Moore, M.K., DiGangi, E.A., Niño Ruíz, F.P., Hidalgo Davila, O.J., Sanabria Medina, C., 2016. Metric sex estimation from the postcranial skeleton for the Colombian population. Forensic. Sci. Int. 262, 286.e1–286.e8.

Murad, T.A., Boddy, M.A., 1987. A case with bear facts. J. Forensic. Sci. 32 (6), 1819–1826.

National Academy of Sciences, 2009. Strengthening Forensic Science in the United States: A Path Forward. National Academies Press, Washington, DC.

Osteoware: Standardized skeletal documentation, 2016. Smithsonian Institution, Washington, DC.

Ousley, S., 2004. 3Skull computer program. Version 2.1.111.

Ousley, S., and Jantz, R., 2005. Fordisc 3.1: Computerized forensic discriminant functions. University of Tennessee, Knoxville.

Peterson, J.L., Hickman, M.J., Strom, K.J., Johnson, D.J., 2013. Effect of forensic evidence on criminal justice case processing. J. Forensic. Sci. 58 (S1), S78–S90.

Pickering, T.R., Carlson, K.J., 2004. Baboon taphonomy and its relevance to the investigation of large felid involvement in human forensic cases. Forensic. Sci. Int. 144 (1), 37–44.

Pokines, J., Symes, S.A. (Eds.), 2014. Manual of Forensic Taphonomy.. CRC Press, Boca Raton.

Pokines, J.T., Sussman, R., Gough, M., Ralston, C., McLeod, E., Brun, K., et al., 2016. Taphonomic analysis of Rodentia and Lagomorpha bone gnawing based upon incisor size. J. Forensic. Sci. 62 (1), 50–66.

President's Council of Advisors on Science and Technology, 2016. Forensic Science in Criminal Courts: Ensuring Scientific Validity of Feature-Comparison Methods. Executive Office of the President, Washington, DC.

Reeves, N.M., 2009. Taphonomic effects of vulture scavenging. J. Forensic. Sci. 54 (3), 523–528.

Rodriguez, W.C., Bass, W.M., 1983. Insect activity and its relationship to decay rates of human cadavers in East Tennessee. J. Forensic. Sci. 28 (2), 423–432.

Rule 702. Testimony by expert witnesses. LII: Cornell University Law School. https://www.law.cornell.edu/rules/fre/rule_702.

Saladié, P., Huguet, R., Díez, C., Rodríguez-Hidalgo, A., Carbonell, E., 2013. Taphonomic modifications produced by modern brown bears (*Ursus arctos*). Int. J. Osteoarchaeol. 23 (1), 13–33.

Schotsmans, E.M., Márquez-Grant, N., Forbes, S.L., 2017. (Eds.), Taphonomy of Human Remains: Forensic Analysis of the Dead and the Depositional Environment.. Wiley-Blackwell, Hoboken.

Shirley, N.R., Ramirez-Montes, P.A., 2015. Age estimation in forensic anthropology: quantification of observer error in phase versus component-based methods. J. Forensic. Sci. 60 (1), 107–111.

Simmons, T., Adlam, R.E., Moffatt, C., 2010. Debugging decomposition data—comparative taphonomic studies and the influence of insects and carcass size on decomposition rate. J. Forensic. Sci. 55 (1), 8–13.

Spradley, M.K., Jantz, R.L., 2011. Sex estimation in forensic anthropology: Skull versus postcranial elements. J. Forensic. Sci. 56 (2), 289–296.

Spradley, M.K., Hamilton, M.D., Giordano, A., 2012. Spatial patterning of vulture scavenged human remains. Forensic. Sci. Int. 219, 57–63.

Spradley, M.K., Anderson, B.E., Tise, M.L., 2015. Postcranial sex estimation criteria for Mexican Hispanics. J. Forensic. Sci. 60 (S1), S27–S31.

Steadman, D.W., Worne, H., 2007. Canine scavenging of human remains in an indoor setting. Forensic. Sci. Int. 173 (1), 78–82.

Steadman, D.W., Adams, B.J., Konigsberg, L.W., 2006. Statistical basis for positive identification in forensic anthropology. Am. J. Phys. Anthropol. 131 (1), 15–26.

Suckling, J.K., Spradley, M.K., Godde, K., 2016. A longitudinal study on human outdoor decomposition in Central Texas. J. Forensic. Sci. 61 (1), 19–25.

Sutherland, A., Myburgh, J., Steyn, M., Becker, P., 2013. The effect of body size on the rate of decomposition in a temperate region of South Africa. Forensic. Sci. Int. 231 (1), 257–262.

Synstelien, J.A., 2013. Raccoon modification of human skeletal remains. *American Journal of Physical Anthropology* 150(S56): 268. Program of the 82nd Annual Meeting of the American Association of Physical Anthropologists, Knoxville.

Synstelien, J.A., Klippel, W.E., 2005. Raccoon (*Procyon lotor*) foraging as a taphonomic agent of soft tissue modification and scene alteration. Proc. Am. Acad. Foren. Sci. 11, 333–334.

Tsokos, M., Schulz, F., 1999. Indoor postmortem animal interference by carnivores and rodents: Report of two cases and review of the literature. Int. J. Legal. Med. 112 (2), 115–119.

Tsokos, M., Matschke, J., Gehl, A., Koops, E., Püschel, K., 1999. Skin and soft tissue artifacts due to postmortem damage caused by rodents. Forensic. Sci. Int. 104 (1), 47–57.

Wellford, C., Cronin, J., 2000. Clearing up homicide clearance rates. Natl Inst. Justice J 243, 2–7. April.

Willey, P., Snyder, L., 1989. Canid modification of human remains: Implications for time-since-death estimations. J. Forensic. Sci. 34 (4), 894–901.

Wright, R., 2008. Detection of likely ancestry using CRANID. In: Oxenham, M. (Ed.), Forensic Approaches to Death, Disaster and Abuse. Australian Academic Press, Bowen Hills, pp. 111–122.

CHAPTER 2

Unwitting Accomplices: Scavengers and Forensic Investigation

INTRODUCTION
SCAVENGING AND IDENTIFICATION
SCAVENGING AND TRAUMA ANALYSIS
SCAVENGING AND POSTMORTEM INTERVAL ESTIMATION
CONCLUSION
REFERENCES

INTRODUCTION

While vertebrate scavenging serves an important function within an ecosystem, it can wreak havoc on death investigations. Vertebrate scavenging damages, disarticulates, and scatters human remains and associated personal effects, disorganizing the death scene by disrupting the depositional context (Moraitis and Spiliopoulou, 2010). Such disorganization reduces the information that can be gleaned from a scene using traditional investigative methods, as critical evidence may be destroyed (e.g., consumption of the finger pads used for fingerprint identifications) or altered (e.g., distortion of perimortem sharp trauma due to targeted feeding at areas of trauma); and pseudo-evidence may be produced (e.g., postmortem damage mistaken for intentional dismemberment) (Pokines and Symes, 2014).

Consequently, all aspects of the scene become more difficult to interpret, from identification of the deceased to understanding circumstances and timing of death. One illustration of this is a case of shark scavenging presented by Anderson et al. (2003), in which two individuals were missing and body parts with presumptive shark bite marks were found on a Hawaiian beach. In this instance, several skeletal elements were absent, either due to sharks or other postmortem processes,

Forensic Taphonomy and Ecology of North American Scavengers. DOI: https://doi.org/10.1016/B978-0-12-813243-2.00002-6

and a combination of experts including forensic anthropologists, forensic entomologists, forensic pathologists, and shark biologists were needed to resolve the case (Anderson et al., 2003).

Another illustration of the complexity introduced by animal scavenging is a report of an indoor death scene with one deceased human and two living dogs (Hart, 2015). The body was discovered just a couple of days after the decedent was last seen alive, and it was apparent that the dogs had interacted with the body at some point, as the head was covered in bite marks. Subsequent anthropological and pathological analysis revealed that the individual had suffered a gunshot wound to the head, and it was determined that the dogs had scavenged the remains after death (Hart, 2015). Had more time elapsed between death and discovery, or a less careful analysis taken place, a different interpretation could have been made: namely, that the dogs had attacked their owner rather than just scavenged his remains.

There is also the possibility that scavenging will impact case resolution itself. A study by Jani and Gupta (2004) found that of the 13 forensic cases involving animal scavenging reported to M.P. Shah Medical College in the year 2000, victims were not able to be identified in 6 cases and cause of death could not be determined in 4 cases. The difficulties presented by animal-scavenged remains are apparent, as discussed throughout this chapter. Unfortunately, Jani and Gupta (2004) did not report corresponding statistics for non-scavenged cases, and so it cannot be determined whether the outcomes of scavenged cases were significantly different from those of other casework conducted at the facility, although this conclusion is implied throughout the article. Despite this, it remains clear that scavengers can confound a death investigation in several ways.

This chapter presents an overview of human identification, trauma analysis, and postmortem interval (PMI) investigation in forensic anthropology while incorporating discussion of how scavengers can hinder or complicate these analyses. Such an understanding is critical if interpretations about death scenes and human skeletal remains are to be accurate. Investigators must be able to positively identify decedents, analyze trauma, and assess how long individuals have been deceased. Knowledge about scavenger behavior and morphology can assist with sorting through the noise scavengers introduce to such analyses.

SCAVENGING AND IDENTIFICATION

Identification of human remains is one of the major end goals of medicolegal investigations. However, identification cannot proceed without some indication as to whom the decedent may be, so that comparisons to extant records and datasets (e.g., DNA, dental records) can occur. This process ranges from relatively straightforward (i.e., one person dies in their own house) to very complex (i.e., the World Trade Center disaster on 9/11 (MacKinnon and Mundorff, 2007) or the Southeast Asian tsunami of 2004 (Wright et al., 2015)).

The complexity of the identification process is determined by the type of information available to investigators from the onset of the case. Closed populations are those where information exists on everyone present, such as a passenger manifest for an aircraft crash (Buck and Briggs, 2009). However, with open populations, the identities of whom was present for the fatal event is unknown from the outset, such as what might occur following a tornado or nightclub fire (Buck and Briggs, 2009). Identifying individuals in closed populations is more straightforward because it requires compiling antemortem data and comparison with the present human remains until everyone is accounted for. Open populations, however, present medicolegal professionals with a more daunting task as an investigation into identity possibilities must occur before comparisons of antemortem and postmortem data can be conducted.

There are legal as well as ethical reasons to identify decedents. Legally, it may be critical for dissolution of an estate or with identifying whom a perpetrator may be. Ethically, it is important to provide relatives and friends with assurance that the remains in the coffin or urn are indeed those of their loved one. While identification may seem relatively straightforward, it raises a complex set of issues related to how certain we *can*, *should*, or *must* be when identifying individuals (Anderson, 2007; Steadman et al., 2006, 2007). Comprehensive discussion of such issues and the subtleties involved are beyond the scope of this chapter, and therefore for our purposes here we present a simplified discussion.

There are two types of identification: positive and presumptive. Positive identification is typically based on some biological characteristic of an individual that can be linked to that person, to the exclusion of (almost) everyone else. When remains are damaged or decomposed beyond facial recognition, DNA or fingerprints typically allow such identifications to be made, as do comparisons of antemortem and postmortem dental data. Less common but still with individualizing possibility would be unique tattoos (Miranda, 2015); surgical sternotomy wires (Fleischman, 2015) or other orthopedic implants (Wilson et al., 2011); or comparison of unique features such as frontal sinus morphology on antemortem and postmortem X-rays (Christensen, 2005; Viner, 2014) among other individualizing characteristics. When the above evidence is absent or unavailable, presumptive, or circumstantial, identifications are made. Items that could establish a presumptive identification could be personal effects found in association with a decedent such as items of clothing, jewelry, cell phone, and/or identity cards. For example, in one case, a presumptive identification was established when a key found in the decedent's pocket unlocked the door to the house where he was staying (Burns, 2002). Depending upon the country and situation, an identification effected in this manner may be enough from a legal standpoint (Baraybar, 2014; DiGangi et al., 2009).

Vertebrate scavenging makes identification of the decedent more difficult. Moderate to severe scavenging renders visual identification in the early postmortem period impossible. Identifying facial features are destroyed when soft tissue is stripped from the face and neck, as is common in cases of indoor scavenging by pet dogs and cats (Byard et al., 2002; Colard et al., 2015; Tsokos et al., 1999). For instance, Jani and Gupta (2004) identified the face and neck to be the most frequently scavenged body parts in their small sample of 13 individuals. Removal and consumption of skin may also distort or destroy individualizing scars and tattoos, while consumption of soft tissue from the digits eliminates fingerprints as a means of identification (Byard et al., 2002). Occasionally, a fortuitous investigator may recover intact digits from a large scavenger's scat or stomach contents, allowing for fingerprint analysis to proceed (Pickering, 2001; Pickering and Carlson, 2004; Rathbun and Rathbun, 1997).

Although vertebrate scavenging significantly complicates the issue, identifications (presumptive or positive, depending on the available data) may even be made in cases of near-complete consumption using remaining personal effects, forensic anthropological analysis, and/or genetics (Steadman and Worne, 2007). For example, Birkby et al. (2008)

and Beatrice and Soler (2016) discuss strategies respectively referred to as the cultural and biocultural profile, used to identify individuals who perish in the Sonoran Desert while crossing the border between Mexico and the United States. Many of these bodies are altered by desert scavengers (Beck et al., 2015). Establishing a biocultural profile is a step toward identification by ascertaining likely citizenship, as it includes observation of a combination of evidence such as dental health, skeletal stress markers, and personal effects.

Forensic anthropological analysis traditionally consists of estimating the biological profile: age-at-death, sex (adults only), ancestry, and stature (for more information about the biological profile, Appendix B contains an abridged bibliography of recommended reading). In recent decades, the biological profile has also incorporated analyses of trauma, taphonomy, and archeological context as discussed in Chapter 7. In addition, forensic anthropologists undertake comparison of antemortem and postmortem skeletal data (e.g., radiographs) for identification purposes. Scavenger destruction will include identifying features as well as those important for other analyses. However, so long as several bones are present and relatively undamaged—especially the ossa coxae (hip bones), facial skeleton, and at least one complete long bone from the arm or leg—all biological profile parameters can be estimated. The second author recently had a case where extensive taphonomic damage had resulted in dozens of bone fragments with almost every bone affected. However, she was able to estimate all parameters except ancestry after carefully piecing the skeleton back together. Despite extensive damage, unique features were still present that would have allowed comparison with antemortem records, had they existed.

DNA is the gold standard for positive identification. However, four issues exist with obtaining DNA from deceased individuals. First, DNA begins to degrade once cellular function ceases. Therefore, the difficulty of obtaining a profile increases with time since death (TSD) (Allentoft et al., 2012). Second, decedent remains can be contaminated by DNA from those who come into contact with them; or analysis can be inhibited by acids in the soil (Matheson et al., 2010). Therefore, body parts that are less likely to be contaminated such as teeth or thick bones (e.g., the femur) are preferred for analysis. Further, specialized ancient DNA laboratories that have extreme contamination-proof protocols in place must be used especially for skeletonized individuals (Cabana et al., 2013). Third, as with any postmortem data, antemortem samples must exist to which they can be compared. If enough time has elapsed since death that no known samples of the decedent's DNA exist (e.g., toothbrush or hairbrush samples), then the DNA sample can be compared to biological relatives. However, the more distant the relationship (e.g., aunt, cousin), the more complex the comparison process. Fourth, DNA extraction and analysis are expensive, with invoices generated for positive as well as negative results.

SCAVENGING AND TRAUMA ANALYSIS

In addition to making identification more difficult, vertebrate scavenging may confound the circumstances surrounding a death. In deaths due to injury, scavenging may modify wounds, as evidence of the wound is frequently consumed along with the adjacent soft tissue. Unfortunately, wounds are attractive entry points for many scavengers, and the tissues surrounding them may be consumed early in the scavenging period (Byard et al., 2002; Haglund, 1997; Willey and Snyder, 1989). In addition, consumption of organs by vertebrate scavengers destroys evidence of natural death, such as organ failure or disease (Byard et al., 2002). However, organs would decay naturally due to decompositional processes, so scavenger interference here may not be as much of a concern as compared to animal interaction with skeletal injury.

When a wound can be evidenced in soft tissue or skeletal remains, scavenging may prevent accurate assessments of its characteristics or origin. For example, in a case described by Puskas (2003), evidence of canid scavenging on skeletal remains obfuscated the origin of perimortem bilateral fractures on the coronoid processes of the mandible. Although scavenging was initially considered as a possible cause of these fractures, the presence of similar fractures in nonscavenged cases and evidence of the victim's intent to commit suicide led to the determination that the fractures were artifacts of intraoral gunshot trauma. Knowing that scavengers modify and disguise trauma, perpetrators may intentionally feed victims to an animal in an attempt to obliterate evidence of criminal activity. In a case described by Boglioli et al. (2000), a father attempted to disguise evidence of child abuse by dismembering and feeding his infant son to the family dog. In this case, the infant's parents reported the death as an attack by the dog, and the dog's scavenging had imitated the perimortem trauma of an animal attack. Upon closer inspection however, tool marks from the initial dismemberment were discovered.

There are three categories of trauma, defined temporally with respect to death. Antemortem trauma (before death) is characterized by evidence of healing. Therefore scavenging damage can be easily discerned from this type of trauma, although scavengers certainly could destroy or alter evidence of it. The second type, perimortem trauma, characterizes trauma that occurs around the time of death. This is the type of trauma that vertebrate scavenging is the most likely to mimic. The third type, postmortem damage, includes all destructive processes that result in breakage, fragmentation, or fractures after death. Therefore, scavenging itself is an agent of postmortem damage. Note that the results of all destructive postmortem processes are referred to as *damage* rather than as *trauma*. *Trauma* refers to injury and therefore it is not appropriate to use this word for postmortem contexts, because dead bones cannot be injured.

Fortunately, distinguishing between perimortem trauma, antemortem trauma, and postmortem damage is possible, although the lines are often blurred when differentiating between perimortem trauma and postmortem damage. Antemortem trauma is the most easily identified as it contains evidence of wound healing in soft tissue and/or bone remodeling (Kondo and Ishida, 2010; Sauer, 1998). In soft tissue injuries, pathologists assess timing by looking for vital responses. Such responses can be measured by looking at levels of wound healing biomarkers, including cytokines and angiogenic growth factors, which can inform (1) whether the individual was alive when the wounds were sustained; and (2) the age of the wound at the time of death (Kondo and Ishida, 2010).

However, in bone, perimortem trauma and postmortem damage are more difficult to distinguish from one another because anthropologists are much more limited with the tools available for perimortem versus postmortem assessments. The reason for this difficulty is that trauma must be examined with respect to the biomechanical response of the bone, given bone's intrinsic properties (Symes et al., 2012). Bones consist of a mixture of collagen and minerals, which give bone tissue its unique ability to be simultaneously elastic and stiff. Fresh bone contains more collagen and thus responds more elastically to applied force than dry bone, in which the collagen has decayed (Sauer, 1998). A major difficulty with trauma analysis lies in the fact that bone collagen does not immediately disappear once death occurs. In fact, bones may retain their collagen content for some time after death. Fracture mechanics and appearance are dictated by bone's composition, so damage occurring after death, prior to the decomposition of bone collagen, may appear fresh (Symes et al., 2012).

Bent or warped bone with radiating or concentric fractures typify perimortem trauma in fresh bone, i.e., bone with the living ratio of collagen to mineral content (Kimmerle and Baraybar, 2008). Dry bone, or bone that has lost the collagen component, tends to have cleaner and straighter fracture lines that are oriented at right angles to the bone's cortex (Cardoso et al., 2015). Other taphonomic indicators may also assist with differentiating between perimortem trauma and postmortem damage. For example, soil staining may result in color differences between a bone's surface and underlying bone more recently exposed by postmortem damage (Ubelaker and Adams, 1995) (Fig. 2.1).

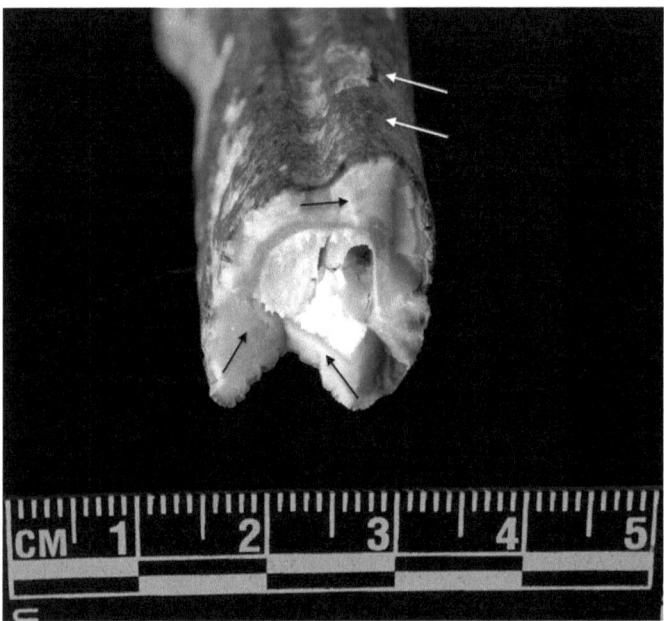

*Figure 2.1 Postmortem damage to a deer (*Cervidae spp.*) metatarsus. Note soil staining on external cortex (white arrows) versus a lighter color internally (black arrows), indicating recent damage. Also note retained marrow.* Photograph by David Tuttle.

Intrinsic, individual characteristics can also play a role in determining the likelihood of bone damage by scavengers. Bone density, or the amount of bone mineral to bone volume, can dictate the likelihood and extent of damage, with low-density trabecular bone being more susceptible to damage than high-density cortical bone (Delaney-Rivera et al., 2009; National Institute of Health, 2008). Bones from juvenile and elderly individuals are particularly prone to damage; the former due to lower bone ossification and incomplete epiphyseal fusion (Ballejo et al., 2016), and the latter due to age-related decline in bone density rates. Bone density is also obviously relevant for the propagation and formation of fractures in the antemortem and perimortem periods, but such a characteristic can disproportionately contribute to scavenger destruction. For example, imagine two skeletons: one from an 80-year old female and the second from a 20-year old male. The male individual has thick and relatively heavy bones, while the female's bones are fragile and light. While this is yet to be tested experimentally, it stands to reason that the fragile bones would be destroyed much more readily and rapidly than the firm ones. In the next section, we discuss how disproportionate destruction can potentially lead to the overestimation of PMI.

Finally, it should be noted that evidence of trauma on a skeletal element is obviously only useful if that element is recovered for analysis. Disarticulation and scattering of remains is common with prolonged exposure to scavengers and this can compromise recovery efforts (Calce and Rogers, 2007), discussed further in Chapter 7.

SCAVENGING AND POSTMORTEM INTERVAL ESTIMATION

The PMI refers to the amount of time that has passed since an individual has died. Also known as time since death (TSD), these two terms are used interchangeably. In forensic casework, the PMI estimate serves multiple purposes by temporally reducing potential victim pools and corroborating (or refuting) suspect alibis. For example, a PMI estimate of 2–3 months in a homicide case would eliminate missing persons who were last seen alive 2 weeks ago, while also establishing that a suspect who is known to have been traveling abroad during that period could not have committed the crime. The PMI estimate thus allows a focusing of the investigation and is consequently critical. Unfortunately, the longer the PMI, the more difficult its accurate estimation tends to be.

Determining TSD for the recently deceased (hours to days) is within the purview of forensic pathologists, who examine soft tissue changes such as body temperature, muscle rigidity, and skin discoloration (Wilson-Taylor, 2013). However, once decomposition has progressed, such indicators are no longer reliable due to tissue breakdown. At this point, other disciplines such as forensic entomology and forensic anthropology have a role to play.

Forensic entomology is the application of insect biology to forensic questions. Insects are well-known colonizers of carcasses and have a substantial effect on decomposition rates, as larvae from blow flies (*Calliphoridae* spp.) in particular eat decomposing soft tissues (Vanin and Huchet, 2017; Wilson-Taylor, 2013). It is possible to determine when a body was first colonized by collecting associated larvae and observing how long they take to develop (pupate) into adult flies, since insects develop at known rates given a certain optimal temperature (Vanin and Huchet, 2017). For individuals who have likely been dead for longer than a few days, insects can still be used to estimate PMI because different species will colonize the body in a predictable order regardless of season or geography (Vanin and Huchet, 2017).

Estimating TSD officially became part of the forensic anthropology repertoire in the 1980s with the founding of the Anthropology Research Facility at the University of Tennessee, Knoxville by Dr. William Bass (Vidoli et al., 2017). As the story famously goes, Dr. Bass was asked to give an opinion on the biological profile and PMI for a set of remains from a disturbed burial. He estimated that the individual had been dead for 6–12 months (Wilson-Taylor, 2013). Unfortunately, the remains turned out to belong to a Confederate Colonel from the American Civil War, which ended in 1865. With this embarrassment behind him, Dr. Bass decided that the question of TSD needed to be scientifically studied, and the world's first human decomposition research facility was born (Vidoli et al., 2017).

Early research involved recording how long bodies took to decompose and qualitative observations about the stages the bodies went through to skeletonization (e.g., Bass, 1997). However, it was gradually acknowledged that the vast number of variables involved, and the interplay between them, made decomposition a difficult process to study. As a result, researchers have attempted to quantify decompositional changes, based on ambient temperature (accumulated degree days, ADD) or morphology (total body score). Given that temperature is the most important variable when it comes to decomposition (Wilson-Taylor, 2013), ADD, first introduced by Vass et al. (1992), is a method that combines ambient temperature and elapsed time. ADD is an important concept because the general statement that "skeletonization took 2 months" is not useful when comparing decomposition processes in southern Florida versus northern Minnesota, for example. However, adding up daily temperatures over time allows for such comparisons (Simmons, 2017).

Megyesi et al. (2005) presented the total body score, which combines the concept of ADD with a point scoring system for different decomposition phases for three body regions (e.g., 3 points for gray/green trunk discoloration, 8 points for trunk mummification, 10 points for limb skeletonization) that result in a final TSD estimate using a regression formula. However, several studies have found that overall, total body score is a poor predictor of PMI due to the variability of the decomposition environment (Dabbs et al., 2016; Marhoff et al., 2016; Suckling et al., 2016; Sutherland et al., 2013). Nevertheless, researchers continue to work on this problem in various environments, proposing the degree-day index (Michaud and Moreau, 2011), degree of decomposition index (Marhoff et al., 2016), and the accumulated decomposition score system (Gleiber et al., 2017).

While scientists continue to conduct research on refining PMI in the entomological and anthropological realms, researchers in other areas have contributed their knowledge and expertise to the problem. For example, studies in botany (e.g., Willey and Heilman, 1987), biochemistry and soil chemistry (e.g., Vass et al., 1992; Vass, 2017), and soil and individual microbiome changes (Damann, 2017) are being conducted to assist with PMI refinement.

In summary, PMI estimation is a popular research topic in anthropology and other disciplines, but there are limitations of the existing methods, largely due to the variables involved. In particular, understanding how scavengers can bias estimates is largely ignored. For example, several vertebrate scavengers (e.g., opossums, raccoons) eat insects and will simultaneously snack on maggots and the remains themselves. Given that maggots are a primary consumer of the remains, reducing their population may serve to delay decomposition (Anderson, 2011; Pechal et al., 2014). In contrast, it is now well known that consumption of soft tissue by vultures can rapidly accelerate decomposition, leading to an overestimation of PMI (Reeves, 2009; Spradley et al., 2012; Suckling et al., 2016). In a more unusual example, remains swallowed whole by tiger sharks (*Galeocerdo cuvieri*) may stay in the stomach for days or even weeks before being digested; how this prolonged exposure to the digestive tract affects the normal rate of decomposition is unknown (Rathbun and Rathbun, 1997).

CONCLUSION

As demonstrated, scavengers can introduce a layer of complexity to forensic cases by destroying evidence, scattering remains, and mimicking perimortem trauma caused by human perpetrators. Their contamination of evidence can also complicate estimation of the PMI. However, comprehensive knowledge of human identification methods, trauma analysis, and estimation of TSD can be used in concert with information about typical scavenger destructive patterns to contribute to case resolution. In the next chapter, we discuss the evolution of scavenging as a foraging strategy to further contextualize the ecology of vertebrate scavengers.

REFERENCES

Allentoft, M.E., Collins, M., Harker, D., Haile, J., Oskam, C.L., Hale, M.L., et al., 2012. The half-life of DNA in bone: measuring decay kinetics in 158 dated fossils. Proc. R. Soc. Lond. B Biol. Sci. 279 (1748), 4724–4733.

Anderson, B., Manoukian, A., Holland, T., Grant, W., 2003. A death in paradise: human remains scavenged by a shark. In: Steadman, D. (Ed.), Hard Evidence: Case Studies in Forensic Anthropology, first ed. Pearson Education, Inc, Upper Saddle River, pp. 186–196.

Anderson, B.E., 2007. Statistical basis for positive identification in forensic anthropology. Am. J. Phys. Anthropol. 133 (1), 741–742.

Anderson, G.S., 2011. Comparison of decomposition rates and faunal colonization of carrion in indoor and outdoor environments. J. Forensic Sci. 56 (1), 136–142.

Ballejo, F., Fernández, F.J., De Santis, L.J.M., Montalvo, C.I., 2016. Taphonomy and dispersion of bones scavenged by New World vultures and caracaras in Northwestern Patagonia: implications for the formation of archaeological sites. Archaeol. Anthropol. Sci. 8 (2), 305–315.

Baraybar, J., 2014. Exhumation, analyse et identification dans les conflits armés: L'examen des victims de Putis, Pérou. In: Delabarde, T., Ludes, B. (Eds.), Manuel Pratique d'anthropologie Médico-Légale. Editions Eksa.

Bass, W.M., 1997. Outdoor decomposition rates in Tennessee. In: Haglund, W.D., Sorg, M.H. (Eds.), Forensic Taphonomy: The Postmortem Fate of Human Remains. CRC Press, Boca Raton, pp. 181–186.

Beatrice, J.S., Soler, A., 2016. Skeletal indicators of stress: a component of the biocultural profile of undocumented migrants in southern Arizona. J. Forensic Sci. 61 (5), 1164–1172.

Beck, J., Ostericher, I., Sollish, G., De Leon, J., 2015. Animal scavenging and scattering and the implications for documenting the deaths of undocumented border crossers in the Sonoran Desert. J. Forensic Sci. 60 (S1), S11–S20.

Birkby, W.H., Fenton, T.W., Anderson, B.E., 2008. Identifying southwest Hispanics using non-metric traits and the cultural profile. J. Forensic Sci. 53 (1), 29–33.

Boglioli, L.R., Taff, M.L., Turkel, S.J., Taylor, J.V., Peterson, C.D., 2000. Unusual infant death-dog attack or postmortem mutilation after child abuse?. Am. J. Forensic Med. Pathol. 21 (4), 389–394.

Buck, A.M., Briggs, C.A., 2009. The role of the anthropologist in disaster victim identification: The Bali incidents of 2002 and 2004. In: Blau, S., Ubelaker, D. (Eds.), Handbook of Forensic Anthropology and Archaeology. Left Coast Press, Walnut Creek, pp. 407–415.

Burns, K., 2002. Challenges of the Haitian courtroom. Proc. Am. Acad. Forensic Sci. 8, 221–222.

Byard, R.W., James, R.A., Gilbert, J.D., 2002. Diagnostic problems associated with cadaveric trauma from animal activity. Am. J. Forensic Med. Pathol. 23 (3), 238–244.

Cabana, G., Hulsey, B., Pack, F., 2013. Molecular methods. In: DiGangi, E., Moore, M. (Eds.), Research Methods in Human Skeletal Biology. Academic Press, San Diego, pp. 449–482.

Calce, S.E.., Rogers, T.L., 2007. Taphonomic changes to blunt force trauma: a preliminary study. J. Forensic Sci. 52 (3), 519–527.

Cardoso, H.F., Puentes, K.S., Coelho, L.F., 2015. Shot and beaten to death? Suspected projectile and blunt force trauma in a case involving an extended period of postmortem water immersion. In: Passalacqua, N., Rainwater, C. (Eds.), Skeletal Trauma Analysis: Case Studies in Context. Wiley-Blackwell, Hoboken, pp. 90–107.

Christensen, A.M., 2005. Testing the reliability of frontal sinuses in positive identification. J. Forensic Sci. 50 (1), 18–22.

Colard, T., Delannoy, Y., Naji, S., Gosset, D., Hartnett, K., Bécart, A., 2015. Specific patterns of canine scavenging in indoor settings. J. Forensic Sci. 60 (2), 495–500.

Dabbs, G.R., Connor, M., Bytheway, J.A., 2016. Interobserver reliability of the total body score system for quantifying human decomposition. J. Forensic Sci. 61 (2), 445–451.

Damann, F., 2017. Bacterial symbionts and taphonomic agents of humans. In: Schotsmans, E., Márquez-Grant, N., Forbes, S. (Eds.), Taphonomy of Human Remains: Forensic Analysis of the Dead and the Depositional Environment. Wiley-Blackwell, Hoboken, pp. 155–166.

Delaney-Rivera, C., Plummer, T.W., Hodgson, J.A., Forrest, F., Hertel, F., Oliver, J.S., 2009. Pits and pitfalls: taxonomic variability and patterning in tooth mark dimensions. J. Archaeol. Sci. 36 (11), 2597–2608.

DiGangi, E.A., Bethard, J.D., Kimmerle, E.H., Konigsberg, L.W., 2009. A new method for estimating age-at-death from the first rib. Am. J. Phys. Anthropol. 138 (2), 164–176.

Fleischman, J.M., 2015. Radiographic identification using midline medical sternotomy wires. J. Forensic Sci. 60 (S1), S3–S10.

Gleiber, D.S., Meckel, L.A., Siegert, C.C., McDaneld, C.P., Pyle, J.A., Wescott, D.J., 2017. Accumulated decomposition score (ADS): an alternative method to total body score (TBS) for quantifying gross morphological change associated with decomposition. Proc. Am. Acad. Forensic Sci. 23, 206–207.

Haglund, W.D., 1997. Dogs and coyotes: postmortem involvement with human remains. In: Haglund, W.D., Sorg, M.H. (Eds.), Forensic Taphonomy: The Postmortem Fate of Human Remains. CRC Press, Boca Raton, pp. 367–382.

Hart, G., 2015. Man's best friend: a case study of ballistics trauma and animal scavenging. In: Passalacqua, N., Rainwater, C. (Eds.), Skeletal Trauma Analysis: Case Studies in Context. Wiley-Blackwell, Hoboken, pp. 108–117.

Jani, C.B., Gupta, B.D., 2004. An autopsy study on medico-legal evaluation of post-mortem scavenging. Med. Sci. Law 44 (2), 121–126.

Kimmerle, E.H., Baraybar, J.P., 2008. Skeletal Trauma: Identification of Injuries Resulting from Human Rights Abuse and Armed Conflict. CRC Press, Boca Raton.

Kondo, T., Ishida, Y., 2010. Molecular pathology of wound healing. Forensic Sci. Int. 203 (1-3), 93–98.

MacKinnon, G., Mundorff, A.Z., 2007. The World Trade Center–September 11, 2001. In: Thompson, T., Black, S. (Eds.), Forensic Human Identification: An Introduction. CRC Press, Boca Raton, pp. 485–499.

Marhoff, S.J., Fahey, P., Forbes, S.L., Green, H., 2016. Estimating post-mortem interval using accumulated degree-days and a degree of decomposition index in Australia: a validation study. Aust. J. Forensic Sci. 48 (1), 24–36.

Matheson, C.D., Gurney, C., Esau, N., Lehto, R., 2010. Assessing PCR inhibition from humic substances. Open Enzym. Inhib. J. 3 (1), 38–45.

Megyesi, M.S., Nawrocki, S.P., Haskell, N.H., 2005. Using accumulated degree-days to estimate the postmortem interval from decomposed human remains. J. Forensic Sci. 50 (3), 618–626.

Michaud, J.P., Moreau, G., 2011. A statistical approach based on accumulated degree-days to predict decomposition-related processes in forensic studies. J. Forensic Sci. 56 (1), 229–232.

Miranda, M.D., 2015. Forensic Analysis of Tattoos and Tattoo Inks. CRC Press, Boca Raton.

Moraitis, K., Spiliopoulou, C., 2010. Forensic implications of carnivore scavenging on human remains recovered from outdoor locations in Greece. J. Forensic Legal Med. 17 (6), 298–303.

National Institute of Health, 2008. Bone density. Medical Subject Headings: US National Library of Medicine.

Pechal, J.L., Benbow, M.E., Crippen, T.L., Tarone, A.M., Tomberlin, J.K., 2014. Delayed insect access alters carrion decomposition and necrophagous insect community assembly. Ecosphere 5 (4), 1–21.

Pickering, T.R., 2001. Carnivore voiding: a taphonomic process with the potential for the deposition of forensic evidence. J. Forensic Sci. 46 (2), 406–411.

Pickering, T.R., Carlson, K.J., 2004. Baboon taphonomy and its relevance to the investigation of large felid involvement in human forensic cases. Forensic Sci. Int. 144 (1), 37–44.

Pokines, J., Symes, S.A. (Eds.), 2014. Manual of Forensic Taphonomy. CRC Press, Boca Raton.

Puskas, C.M., 2003. Bilateral fractures of the coronoid processes: differential diagnosis of intra-oral gunshot trauma and scavenging using a sheep crania model. J. Forensic Sci. 48 (6), 1219–1225.

Rathbun, T., Rathbun, B., 1997. Human remains recovered from a shark's stomach in South Carolina. In: Haglund, W.D., Sorg, M.H. (Eds.), Forensic Taphonomy: The Postmortem Fate of Human Remains. CRC Press, Boca Raton, pp. 449–456.

Reeves, N.M., 2009. Taphonomic effects of vulture scavenging. J. Forensic Sci. 54 (3), 523–528.

Sauer, N.J., 1998. The timing of injuries and manner of death: distinguishing among antemortem, perimortem and postmortem trauma. In: Reichs, K.J. (Ed.), Forensic Osteology: Advances in the Identification of Human Remains.. Charles C. Thomas, Springfield, pp. 321–332.

Simmons, T., 2017. Post-mortem interval estimation: an overview of techniques. In: Schotsmans, E., Márquez-Grant, N., Forbes, S. (Eds.), Taphonomy of Human Remains: Forensic Analysis of the Dead and the Depositional Environment. Wiley-Blackwell, Hoboken, pp. 134–142.

Spradley, M.K., Hamilton, M.D., Giordano, A., 2012. Spatial patterning of vulture scavenged human remains. Forensic Sci. Int. 219 (1–3), 57–63.

Steadman, D.W., Worne, H., 2007. Canine scavenging of human remains in an indoor setting. Forensic Sci. Int. 173 (1), 78−82.

Steadman, D.W., Adams, B.J., Konigsberg, L.W., 2006. Statistical basis for positive identification in forensic anthropology. Am. J. Phys. Anthropol. 131 (1), 15−26.

Steadman, D.W., Adams, B.J., Konigsberg, L.W., 2007. Statistical basis for positive identification in forensic anthropology: response to Anderson. Am. J. Phys. Anthropol. 133 (1), 741−742.

Suckling, J.K., Spradley, M.K., Godde, K., 2016. A longitudinal study on human outdoor decomposition in Central Texas. J. Forensic Sci. 61 (1), 19−25.

Sutherland, A., Myburgh, J., Steyn, M., Becker, P., 2013. The effect of body size on the rate of decomposition in a temperate region of South Africa. Forensic Sci. Int. 231 (1), 257−262.

Symes, S.A., L'Abbé, E.N., Chapman, E.N., Wolff, I., Dirkmaat, D.C., 2012. Interpreting traumatic injury to bone in medicolegal investigations. In: Dirkmaat, D. (Ed.), A Companion to Forensic Anthropology. Wiley-Blackwell, Hoboken, pp. 340−389.

Tsokos, M., Schulz, F., Püschel, K., 1999. Unusual injury pattern in a case of postmortem animal depredation by a domestic German Shepherd. Am. J. Forensic Med. Pathol. 20 (3), 247−250.

Ubelaker, D.H., Adams, B.J., 1995. Differentiation of perimortem and postmortem trauma using taphonomic indicators. J. Forensic Sci. 40 (3), 509−512.

Vanin, S., Huchet, J.-B., 2017. Forensic entomology and funerary archaeoentomology. In: Schotsmans, E., Márquez-Grant, N., Forbes, S. (Eds.), Taphonomy of Human Remains: Forensic Analysis of the Dead and the Depositional Environment. Wiley-Blackwell, Hoboken, pp. 167−186.

Vass, A., 2017. The use of volatile fatty acid biomarkers to estimate the post-mortem interval. In: Schotsmans, E., Márquez-Grant, N., Forbes, S. (Eds.), Taphonomy of Human Remains: Forensic Analysis of the Dead and the Depositional Environment. Wiley-Blackwell, Hoboken, pp. 387−393.

Vass, A.A., Bass, W.M., Wolt, J.D., Foss, J.E., Ammons, J.T., 1992. Time since death determinations of human cadavers using soil solution. J. Forensic Sci. 37 (5), 1236−1253.

Vidoli, G.M., Steadman, D.W., Devlin, J.B., Jantz, L.M., 2017. History and development of the first Anthropology Research Facility, Knoxville, Tennessee. In: Schotsmans, E., Márquez-Grant, N., Forbes, S. (Eds.), Taphonomy of Human Remains: Forensic Analysis of the Dead and the Depositional Environment. Wiley-Blackwell, Hoboken, pp. 463−475.

Viner, M., 2014. The use of radiology in mass fatality events. In: Adams, B., Byrd, J. (Eds.), Commingled Human Remains: Methods in Recovery, Analysis, and Identification. Academic Press, San Diego, pp. 87−122.

Willey, P., Heilman, A., 1987. Estimating time since death using plant roots and stems. J. Forensic Sci. 32 (5), 1264−1270.

Willey, P., Snyder, L., 1989. Canid modification of human remains: implications for time-since-death estimations. J. Forensic Sci. 34 (4), 894−901.

Wilson, R.J., Bethard, J.D., DiGangi, E.A., 2011. The use of orthopedic surgical devices for forensic identification. J. Forensic Sci. 56 (2), 460−469.

Wilson-Taylor, R., 2013. Time since death estimation and bone weathering. In: Tersigni-Tarrant, M.T., Shirley, N.R. (Eds.), Forensic Anthropology: An Introduction. CRC Press, Boca Raton, pp. 339−380.

Wright, K., Mundorff, A., Chaseling, J., Forrest, A., Maguire, C., Crane, D.I., 2015. A new disaster victim identification management strategy targeting "near identification-threshold" cases: experiences from the Boxing Day tsunami. Forensic Sci. Int. 250, 91−97.

CHAPTER 3

There Is No Such Thing as a Free Lunch: The Evolution of Scavenging

INTRODUCTION

Scavenging is a foraging strategy in which dead and decaying flesh and bone, known as carrion, is consumed to meet nutritional needs. Scavenging is prevalent throughout the entire animal kingdom, although many human cultures worldwide have singled out specific scavengers as being taboo animals, such as rodents, vultures, and crows. However, almost any animal that incorporates meat in its diet, from large carnivorous predators such as the cougar to smaller omnivorous species such as the opossum, will feed from carrion given the opportunity. In fact, research in evolutionary anthropology has long suggested that the practice played a very significant role in the evolution of our species (Blumenschine et al., 1987, 2007; Domínguez-Rodrigo and Barba, 2006, 2007; Shipman, 1986; Speth, 1989), although scavenging by humans is currently considered taboo by westernized populations. As consumption of carrion poses risks for digestive upset and disease transmission, one may wonder how such a foraging strategy could have evolved for any vertebrate animals. This chapter explores the evolution of scavenging behavior and introduces several specific physical and physiological adaptations which improve scavenging efficiency.

Forensic Taphonomy and Ecology of North American Scavengers. DOI: https://doi.org/10.1016/B978-0-12-813243-2.00003-8

SCAVENGING AND ECOLOGY

Scavengers make important, although historically underestimated, contributions to the ecosystem. Scavenging accelerates nutrient cycling, disperses concentrated nutrients, and dilutes pathogenic organisms associated with decomposition (Selva et al., 2005). It transfers significant amounts of energy between trophic levels (i.e., positions within a food web; for example, herbivores or carnivores). In fact, scavengers and other decomposers occupy the dominant trophic position when they are included in food web analyses (Wilson and Wolkovich, 2011). Utilization of available carrion bolsters scavenger populations and has subsequent influences on populations of their prey, predators, competitors, and parasites (Ostfeld and Keesing, 2000).

Not only is carrion a high-quality energy resource, but it requires relatively little energy expenditure when compared to a hunting strategy (DeVault et al., 2003; Houston et al., 1979). The result is a high net energy gain. Because of these extraordinary energetic benefits of scavenging, most vertebrates will scavenge if presented with carrion (Selva and Fortuna, 2007). Opportunistic, or facultative, scavengers will utilize carrion as it is available, and while it is generally abundant, it is obviously only useful if it can be found and consumed. In the absence of carrion, facultative scavengers will rely on their primary feeding strategy (e.g., predation or herbivory) to meet their nutritional needs (DeVault et al., 2003; Wilson and Wolkovich, 2011).

Carrion occurs in resource pulses; it is available in abundance immediately after organisms die but quickly becomes depleted (Ostfeld and Keesing, 2000). Carrion is typically scattered in isolated patches throughout a territory, and the accumulating toxic byproducts of decomposition ensure that a carcass is only edible by vertebrates for a limited time after death, what we refer to in this text as the "vertebrate scavenging window" (DeVault et al., 2003; Houston et al., 1979). High competition for carrion exists, as it is a food resource for microorganisms, invertebrates, and vertebrates alike. Further, the amount of available carrion fluctuates with respect to the current vulnerability of animal populations. A number of ecological variables including climate, community composition, disease transmission, and human activity impact a given population's mortality rate and therefore the availability of carrion within the respective ecosystem (Wilson and Wolkovich, 2011). Here, we discuss how the energetic benefits of scavenging are maximized by adaptations leading to improved foraging efficiency, increased scavenger body size, and reduced competition with other organisms.

SCAVENGING AND EVOLUTION

Due to the unpredictable nature of carrion, obligate scavengers (i.e., specialists whose feeding strategy is carrion-only) must be able to travel greater distances in search of meals and feed less frequently (Houston et al., 1979; Kane et al., 2017). If locomotion is inefficient for long distance travel, the energetic demands of searching for carrion can quickly surpass the energy ultimately gained and the scavenger is at greater risk of starvation (DeVault et al., 2003). Therefore avian species are often identified as well adapted for scavenging, as the low-energy requirements of flight ensure that large areas can be searched for carrion efficiently (Houston et al., 1979; Kane et al., 2017). Some scholars suggest that the superior efficiency of avian scavenging has prevented the persistence of obligate terrestrial scavengers (Ruxton and Houston, 2004).

In the air, visual cues indicating that a carcass has been detected will not be obscured by elevation changes or vegetation, enabling birds to effectively employ group foraging over significant distances. Vultures in the genus *Gyps*, for example, will lower their feet to create drag when descending to a carcass (Dermody et al., 2011). This signal relays information to conspecifics via local enhancement (i.e., drawing attention to the descending vulture's location) to recruit them to the carcass (Cortés-Avizanda et al., 2014). Communal roosting, common among vultures and corvids (the taxonomic family that includes magpies, crows, and ravens), is a behavior that is also argued to enhance group foraging. Dermody et al. (2011) suggest that communal roosts function by allowing birds to start the foraging day within visual range of conspecifics, enabling them to utilize social information earlier in the day and increasing their chances of locating and monopolizing newly available carrion (Fig. 3.1).

Because of carrion's sporadic availability, the high search costs associated with scavenging in terrestrial species have largely thwarted the evolution of obligate scavengers. Vultures, an avian species, are the only extant obligate scavengers, having traded their hunting ability for

*Figure 3.1 Large committee of American black vultures (*Coragyps atratus*) waiting to feed in a nearby tree after their breakfast is interrupted by daily activities at the Forensic Anthropology Research Facility at Texas State University.* Photograph by R.T. Kramer.

two specialized scavenging adaptations: (1) large body size, and (2) reliance on thermal soaring locomotion (Ruxton and Houston, 2004). Large body size promotes survival between meals by increasing energy reserves stored in the body, allowing bigger meals to be taken when carrion is located, and enhancing the ability to defend or usurp carrion resources from other scavengers (Ruxton and Houston, 2004). Increases in body size also correlate with increased visual acuity, allowing detection of carrion from greater distances, including from the air (Kiltie, 2000).

However, if body size increases too much, locomotion may become too energetically expensive for an obligate scavenger to meet their nutritional needs. To combat this problem, vultures rely on thermal soaring. Thermal soaring requires less energy than either terrestrial movement or powered flying, exploiting air currents to gain altitude before gliding for forward and downward movement (Houston et al., 1979; Kane et al., 2017; Ruxton and Houston, 2004). Together, these adaptations improve foraging efficiency, sufficiently increasing net energy gains to allow vultures to survive on carrion alone.

Although birds tend to be better adapted at locating carrion, terrestrial species can successfully compete for a carcass if they can quickly locate and defend it (DeVault et al., 2003). Therefore the most successful terrestrial scavenging adaptations promote efficient locomotion and rapid carcass detection. Conversely, animals unable to expeditiously locate carrion would not be expected to be successful scavengers. For example, felids are less likely than other carnivores to incorporate large proportions of carrion into their diet (Pereira et al., 2014). With the exception of the African cheetah (*Acinonyx jubatus*), most felid species specialize in ambush predation that employs sit-and-wait or stalk-and-pounce hunting behaviors (Williams et al., 2014). An ambush predator's strategy has a lower energetic cost because they sit and wait for prey to approach, as opposed to spending energy actively searching for it (Williams et al., 2014). As discussed earlier, carrion exploitation does not require much energy expenditure beyond initial location and defense of the carcass. Since ambush predators already have a low energetic cost associated with how they obtain food, exploiting carrion would not add much benefit. Further, carrion is less likely to be located by felids since ambush predators tend to cover less ground within their territories on a daily basis (Husseman et al., 2003).

Terrestrial scavenging species benefit from larger home ranges as it reduces resource loss due to prey migrations (Houston et al., 1979). Nonterritorial species therefore have an automatic advantage over territorial scavengers, as nonterritorial species may continually associate with prey that may otherwise migrate out of range, removing prey and carrion resources in the process (Houston et al., 1979). Polar bears (*Ursus maritimus*), for example, prey predominately on highly mobile populations of aquatic mammals and do not exhibit territorial behavior (Pilfold et al., 2014). The nonterritorial behavior of polar bears allows them to gather at high-quality feeding sites, in association with prey populations and the carrion generated by natural or accidental deaths.

Terrestrial scavengers may also benefit from enhanced olfaction, which allows for detection of carrion earlier during decomposition and from greater distances. In general, increased surface area of the olfactory turbinals (or turbinates) in the nasal passages directly correlates with olfactory sensitivity (Green et al., 2012). Canids and other scavenging species with large home ranges, such as the wolverine (*Gulo gulo*), have been found to have a high turbinate surface area (Green et al., 2012). Although this finding could be attributed to the need to locate widely dispersed live prey, larger olfactory turbinals also serve to improve efficiency of carrion detection and location. Incidentally, humans have long taken advantage of the olfactory power of domestic dogs and have trained them to alert a handler upon the detection of many different substances, among them the odor of decomposing human remains. Such animals are known as cadaver dogs, and are used by law enforcement and civilian groups alike to locate human bodies after disaster events or suspected criminal activity (Rebmann et al., 2000).

Instead of improving carrion detection time, a species could develop physiological mechanisms to combat toxins produced by bacteria during decomposition. This extends the time after death that carrion is edible and thus allows these species to inhabit a less crowded niche (i.e., consume carrion that cannot be exploited by competitors) (DeVault et al., 2003). There has been little research into this strategy, although it is suspected that some snake species, which are unable to outcompete highly mobile mammals and birds for fresh carrion, have evolved a tolerance to decomposed carrion (Shivik, 2006). Recent research has also addressed the tolerance of New World vultures to microbial toxins and their apparent immunity to the pathogenic bacteria associated with decomposition. The gastrointestinal tract of these species is highly acidic, effectively eliminating the majority of ingested pathogenic bacteria (Roggenbuck et al., 2014). The hindgut microbiome of vultures is dominated by Clostridia and Fusobacteria species, which contribute to the breakdown of the vultures' decomposing meals. With the help of naturally occurring antibodies, vultures are able to neutralize the toxins produced by the various bacteria present in their meals, including botulinum (the neurotoxin that causes botulism) (Ohishi et al., 1979). These antibodies have also been observed in select members of other carrion-consuming species, including crows, coyotes, and rats (Ohishi et al., 1979).

Caching, or the intentional concealment of food resources, is another behavior that serves to maximize the energetic benefits which can be obtained from carrion. In addition to hiding food, caching also takes the form of parents delivering food to relatively immature offspring in the den, burrow, or nest, a common behavior in many disparate taxa (e.g., birds, canids) (Smith and Reichman, 1984). Caching is more common at higher latitudes, where annual fluctuations in environmental conditions result in seasonal differences in resource availability (Smith and Reichman, 1984). Caching food items, such as seeds or animal flesh, reduces both intra- and interspecific competition for the resource and thus promotes survival during periods of food shortage. This practice is particularly useful for scavengers, as caching often serves to decelerate the decomposition process, therefore preventing resource depletion by microorganisms and invertebrates (Bischoff-Mattson and Mattson, 2009; Smith and Reichman, 1984). There are two types of caching behavior: (1) scatter hoarding, where an animal creates multiple small caches, and (2) larder hoarding, where an animal stores large amounts of food in a single location (Smith and Reichman, 1984). While scatter hoarding is preferred by species that cannot easily monopolize an entire carcass, such as foxes or crows, larder hoarding is utilized by larger species, such as bears, that are able to defend caches from other scavengers.

CONCLUSION

In summary, this chapter has discussed three evolutionary pathways that maximize the energetic benefits of scavenging carrion: (1) improvement of foraging efficiency by reducing the costs of searching for carrion, (2) increasing body size in order to monopolize or usurp carrion once located, and (3) reduction of competition with invertebrates or microorganisms by delaying decomposition (i.e., carcass relocation and caching) or developing a physiological resistance to microbial toxins. Knowledge of the evolved mechanisms that lead to scavenger efficiency (or inefficiency) ultimately sets the context for understanding how morphological, physiological, and behavioral traits of scavengers contribute to the taphonomic signatures they leave behind at a death scene, to be discussed throughout the rest of this volume.

REFERENCES

Bischoff-Mattson, Z., Mattson, D., 2009. Effects of simulated mountain lion caching on decomposition of ungulate carcasses. West. N. Am. Nat. 69 (3), 343–350.

Blumenschine, R.J., Bunn, H.T., Geist, V., Ikawa-Smith, F., Marean, C.W., Payne, A.G., et al., 1987. Characteristics of an early hominid scavenging niche [and comments and reply]. Curr. Anthropol. 28 (4), 383–407.

Blumenschine, R.J., Prassack, K.A., Kreger, C.D., Pante, M.C., 2007. Carnivore tooth-marks, microbial bioerosion, and the invalidation of test of Oldowan hominin scavenging behavior. J. Hum. Evol. 53 (4), 420–426.

Cortés-Avizanda, A., Jovani, R., Donázar, J.A., Grimm, V., 2014. Bird sky networks: how do avian scavengers use social information to find carrion? Ecology 95 (7), 1799–1808.

Dermody, B.J., Tanner, C.J., Jackson, A.L., 2011. The evolutionary pathway to obligate scavenging in Gyps vultures. PLoS One 6 (9), 11–15.

DeVault, T.L., Rhodes, O.E., Shivik, J.A., 2003. Scavenging by vertebrates: behavioral, ecological, and evolutionary perspectives on an important energy transfer pathway in terrestrial ecosystems. Oikos 102 (2), 225–234.

Domínguez-Rodrigo, M., Barba, R., 2006. New estimates of tooth mark and percussion mark frequencies at the FLK Zinj site: the carnivore-hominid-carnivore hypothesis falsified. J. Hum. Evol. 50 (2), 170–194.

Domínguez-Rodrigo, M., Barba, R., 2007. Five more arguments to invalidate the passive scavenging version of the carnivore-hominid-carnivore model: a reply to Blumenschine et al. J. Hum. Evol. 53 (4), 427–433.

Green, P.A., Valkenburgh, B., Pang, B., Bird, D., Rowe, T., Curtis, A., 2012. Respiratory and olfactory turbinal size in canid and arctoid carnivorans. J. Anat. 221 (6), 609–621.

Houston, D.C., Sinclair, A.R.E., Norton-Griffiths, M., 1979. The adaptions of scavengers. In: Sinclair, A.R.E., Norton-Griffiths, M. (Eds.), Serengeti, Dynamics of an Ecosystem. University of Chicago Press, Chicago, pp. 263–286.

Husseman, J.S., Murray, D.L., Power, G., Mack, C., Wenger, C.R., Quigley, H., 2003. Assessing differential prey selection patterns between two sympatric large carnivores. Oikos 101 (3), 591–601.

Kane, A., Healy, K., Guillerme, T., Ruxton, G.D., Jackson, A.L., 2017. A recipe for scavenging in vertebrates-the natural history of a behaviour. Ecography 40 (2), 324–334.

Kiltie, R.A., 2000. Scaling of visual acuity with body size in mammals and birds. Funct. Ecol. 14 (2), 226–234.

Ohishi, I., Sakaguchi, G., Riemann, H., Behymer, D., Hurvell, B., 1979. Antibodies to Clostridium botulinum toxins in free-living birds and mammals. J. Wildl. Dis. 15 (1), 3–9.

Ostfeld, R.S., Keesing, F., 2000. Pulsed resources and community dynamics of consumers in terrestrial ecosystems. Trends Ecol. Evol. 15 (6), 232–237.

Pereira, L.M., Owen-Smith, N., Moleon, M., 2014. Facultative predation and scavenging by mammalian carnivores: seasonal, regional and intra-guild comparisons. Mammal Rev. 44 (1), 44–55.

Pilfold, N.W., Derocher, A.E., Richardson, E., 2014. Influence of intraspecific competition on the distribution of a wide-ranging, non-territorial carnivore. Glob. Ecol. Biogeogr. 23 (4), 425–435.

Rebmann, A., David, E., Sorg, M.H., 2000. Cadaver Dog Handbook: Forensic Training and Tactics for the Recovery of Human Remains. CRC Press, Boca Raton.

Roggenbuck, M., Schnell, I.B., Blom, N., Bælum, J., Bertelsen, M.F., Sicheritz-Pontén, T., et al., 2014. The microbiome of New World vultures. Nat. Commun. 5, 5498.

Ruxton, G.D., Houston, D.C., 2004. Obligate vertebrate scavengers must be large soaring fliers. J. Theoret. Biol. 228 (3), 431–436.

Selva, N., Fortuna, M.A., 2007. The nested structure of a scavenger community. Proc. R. Soc. B Biol. Sci. 274 (1613), 1101–1108.

Selva, N., Jędrzejewska, B., Jędrzejewski, W., Wajrak, A., 2005. Factors affecting carcass use by a guild of scavengers in European temperate woodland. Can. J. Zool. 83 (12), 1590–1601.

Shipman, P., 1986. Scavenging or hunting in early hominids: theoretical framework and tests. Am. Anthropol. 88 (1), 27–43.

Shivik, J.A., 2006. Are vultures birds, and do snakes have venom, because of macro- and micro-scavenger conflict? BioScience 56 (10), 819–823.

Smith, C., Reichman, O., 1984. The evolution of food caching by birds and mammals. Annu. Rev. Ecol. Syst. 15 (1), 329–351.

Speth, J.D., 1989. Early hominid hunting and scavenging: the role of meat as an energy source. J. Hum. Evol. 18 (4), 329–343.

Williams, T.M., Wolfe, L., Davis, T., Kendall, T., Richter, B., Wang, Y., et al., 2014. Instantaneous energetics of puma kills reveal advantage of felid sneak attacks. Science 346 (6205), 81–85.

Wilson, E.E., Wolkovich, E.M., 2011. Scavenging: how carnivores and carrion structure communities. Trends Ecol. Evol. 26 (3), 129–135.

CHAPTER 4

Scavenger Identification Strategies: Interpreting Taphonomic Signatures

INTRODUCTION
FAUNAL AND GENETIC EVIDENCE
BITE MARK EVIDENCE
INTRODUCING THE MAKER'S MARK: TAPHONOMIC SIGNATURES
SOFT TISSUE AND BONE MODIFICATION
PATTERNS OF TISSUE CONSUMPTION
DISARTICULATION, SCATTERING, AND ELEMENT REMOVAL
CONCLUSION
REFERENCES

INTRODUCTION

If investigation strategies are to be informed by animal behavior, is it essential to first identify the scavenger—or scavengers—responsible for any postmortem damage. There are many strategies available to investigators seeking to make such an identification. Oftentimes, faunal evidence—such as scat, feathers or fur, or animal tracks—may be left at the scene. Occasionally, scavenger identification efforts may be aided by DNA amplified from biological evidence, such as saliva left inside a wound. Although its use is controversial, a careful and thorough analysis of bite mark evidence may also contribute to scavenger identification. If biological forms of evidence are lacking, scavenger identity may be inferred through an analysis of taphonomic signature features including characteristic soft tissue damage or bone modification, patterns of disarticulation and scattering, and survivorship of skeletal

Forensic Taphonomy and Ecology of North American Scavengers. DOI: https://doi.org/10.1016/B978-0-12-813243-2.00004-X

elements. Each form of evidence provides information as well as inter-pretative challenges, and the analysis of any one form is inconclusive if presented on its own. An accurate conclusion can be supported by the holistic analysis of multiple types of evidence. This chapter briefly introduces more traditional forms of evidence used for scavenger iden-tifications, including faunal, genetic, and bite mark evidence. We con-clude with a discussion of how taphonomic signatures can be linked to the morphological, physiological, and behavioral characteristics of dif-ferent scavenging taxa.

FAUNAL AND GENETIC EVIDENCE

One method used to identify a scavenger is the analysis of faunal evi-dence (Figs. 4.1 and 4.2). Such evidence includes animal tracks, scat, feathers, fur, or hair, and can be identified with the assistance of a wildlife biologist or related professional (Murad, 1997). Numerous field guides are available for general visual identifications of tracks and scat for North American species (including Elbroch, 2003; Elbroch and Marks, 2001; Halfpenny, 2015a,b; Murie and Elbroch,

*Figure 4.1 Tracks from raccoon (*Procyon lotor*) front paws left in the mud at the Forensic Anthropology Research Facility at Texas State University. Note the prominent claws. Raccoon forepaws are about 4−7 cm in length. Photograph by Jessica C. Galea.*

*Figure 4.2 Dried vulture (*Cathartidae *spp.) track left in the mud at the Forensic Anthropology Research Facility at Texas State University. Track is approximately 7–10 cm long. The schematic in the upper left corner illustrates the basic form of a vulture foot.* Photograph by R.T. Kramer.

2005). Scat is particularly useful: the form and consistency can yield broad taxonomic classifications of its originator, the scavenger's genetic material can be extracted from intestinal cells shed during excretion, and the scat itself may hold bones or bone fragments that testifies to the animal's involvement (Foran et al., 1997a; Gilmour and Skinner, 2012; Murad, 1997; Pickering, 2001).

While macroscopic characteristics of feathers, fur, and hair left at the scene may narrow the pool of potential scavengers (Fig. 4.3), forensic research has tended to focus on microscopic discrimination of animal fibers using measurements and micromorphological traits such as cuticle scale patterns, medulla form, morphology of feather nodes, and distributions of pigments (Dove and Koch, 2011; Sato et al., 2010; Sessions et al., 2009). Such shed materials also lend themselves to genetic identification of scavenging species, including analysis of either DNA or RNA (Foran et al., 1997b; Hogan et al., 2008; Melton and Holland, 2007; Rudnick et al., 2007; Sato et al., 2010; Tsai et al., 2007). Finally, genetic

*Figure 4.3 Domestic dog (*Canis familiaris*) fur inadvertently snagged on a branch in the woods.* Photograph by David Tuttle.

identification of a scavenger species—and sometimes individual animals involved—may also be made from salivary samples taken from wound margins, although the utility of this technique is limited by the quantity and quality of DNA available at the time the death is investigated (Farley et al., 2014; Schulz et al., 2006).

It should be noted that identifications made based on faunal evidence alone may not be accurate. Generally, faunal evidence only demonstrates that the species was present at the scene before discovery—without indicating that the species actually fed from, or even interacted with, the remains (Murad, 1997). Exceptions, of course, include identifications derived from scat that incorporates bone fragments from the decedent, genetic identifications sampled from saliva present inside a wound, or in the fortuitous circumstance that an identifiable tooth becomes dislodged within the remains (Lowry et al., 2009). Anderson et al. (2003) present a case where the latter occurred, with a shark tooth fragment lodged in a human femur.

BITE MARK EVIDENCE

Bite marks have a considerable (and contentious) history of use in forensic science, most often with the intention of matching them to human perpetrators. The validity of bite mark analysis has long been a subject of debate, with scrutiny intensifying in recent decades following U.S. governmental inquiries into forensic methodology, discussed in the introduction of this book (Bowers and Pretty, 2009; Clement and Blackwell, 2010; Page et al., 2012; Pretty and Sweet, 2001; Pretty, 2006). Less frequently, however, bite mark analysis has been employed to identify offending animals in cases of animal-inflicted injury or death (Bernitz et al., 2012; Fonseca et al., 2015; Rollins and Spencer, 1995; Souviron, 2011; Young et al., 2015). Given the right circumstances, such analyses may also be useful in interpreting bite marks produced during vertebrate scavenging.

For instance, the shape of the anterior jaw may be reflected in bite marks in soft tissue. Generally, the jaws of canids are steeply angled with significant curvature of the anterior dentition, the anterior jaws of felids are more square, and the jaws of bears (and the wolverine, *Gulo gulo*) have an intermediate form (Murmann et al., 2006) (Fig. 4.4). Alligators (*Alligator mississippiensis*) have rounded jaws while crocodiles (*Crocodylus acutus*) have triangular jaws with fewer teeth, and there is considerable variation in jaw morphology and

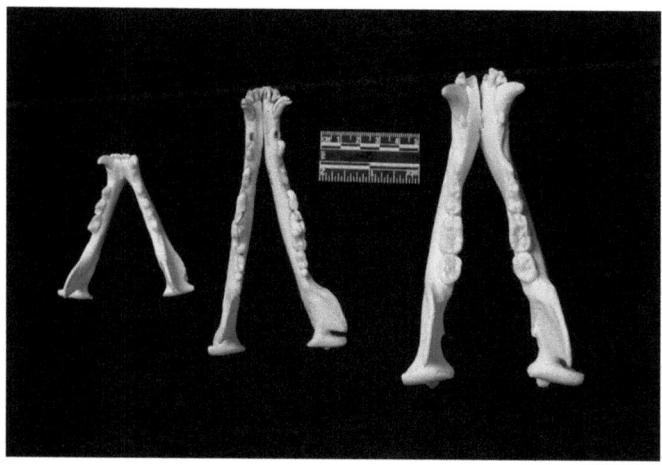

Figure 4.4 From left *to* right: *Mandibles from a bobcat (*Lynx rufus*); coyote (*Canis latrans*); and American black bear (*Ursus americanus*). Note the steep curvature of the anterior dentition of the coyote, more obtuse curvature on the bear's anterior teeth, and squared anterior tooth row of the bobcat.* Photograph by David Tuttle.

dimensions between different species of shark (Harding and Wolf, 2006; Lowry et al., 2009). In skin, jaw shapes may be discernable as arched bruises or lines of punctures, although interpretations should be made with caution as there is high potential for distortion resulting from biting movements, variation in skin elasticity, post-bite decomposition, or attempted preservation techniques (Bush et al., 2010; Rothwell and Thien, 2001; Sheasby and MacDonald, 2001). A number of digital reconstruction methods have been proposed to account for distortion in human bite marks. Such techniques may be adapted for use in interpretations of bites produced by animals, although the details of these methods are outside the scope of this text (see Bowers and Johansen, 2002; Martin-de las Heras et al., 2005; Santoro et al., 2011a; Thali et al., 2003).

Apart from general jaw morphology, dental dimensions may shed light on the species (or even breed) responsible for a bite mark. Murmann et al. (2006) used a zoological collection of nearly 500 animal skulls to collect dental measurements of the maxillary and mandibular canines. The distinctive morphology of canines makes them likely to be reflected in bite marks as pronounced puncture wounds. The resulting publication includes extensive data on the intercanine distance (measured from the tip of the left upper or lower canine to the tip of the corresponding right canine) and canine widths of wild and domestic canids and felids, as well as two bear species and the North American wolverine (Murmann et al., 2006). These data ultimately suggest that while dental measurements and analysis of jaw shape may assist with identification of a broad taxa, there is too much dimensional overlap within carnivore families to distinguish between individual species or breeds from this information alone (Murmann et al., 2006).

However, given a known pool of animal suspects, bite mark analysis may be used to make tentative identifications of responsible individual animals. For example, Santoro et al. (2011b) report the case of a man who was attacked by a pack of stray dogs following a suicide attempt. Five previously reported aggressive dogs which frequented the military base where the victim was discovered were captured and sedated while casts of their dentition were made. Referencing intercanine and interincisal[a] distances, positive matches were made between

---

[a]Interincisal distance is measured from the mesial edge of the left upper central incisor to the mesial edge of the right upper central incisor. Mesial is the anatomical term referring to the part of the tooth closest to the midline of the body (its opposite is distal).

three of the casts and bite marks excised from the remains (Santoro et al., 2011b). The remaining two casts could not be reliably matched due to heavily worn or missing teeth.

The analysis of bite marks, whether produced by humans or animals, remains a controversial forensic method. To ensure the admissibility of evidence in court, it is strongly recommended that investigators seek the advisement of a board-certified forensic odontologist when determining whether a bite mark was produced by a human or nonhuman agent and evaluating what individual or species may be responsible. Given the dearth of research in nonhuman bite mark analysis, such evidence should be interpreted cautiously with heavier consideration given to other available forms of evidence.

INTRODUCING THE MAKER'S MARK: TAPHONOMIC SIGNATURES

Faunal, genetic, and bite mark evidence represent conventional approaches to scavenger identification. These forms of evidence are relatively straightforward to collect and analyze and, at first glance, seem to have neat and tidy interpretations. However, there are limitations to each. Faunal evidence only represents the presence of a species at a death scene and does not necessarily indicate that the species was an active scavenger; genetic evidence may be too degraded to yield an accurate identification (if it is present at all); and the reliability of bite mark evidence is greatly diminished by tissue distortion. Fortunately, there are other—albeit more complex—forms of evidence available to assist with scavenger identification, referred to collectively as the scavenger's taphonomic signature (Marden et al., 2013).

A taphonomic signature describes characteristic damage to human remains that is a product of scavenging by a particular group of organisms, usually at the level of the species (e.g., raccoons) or family (e.g., canids or Canidae). Elements that make up a taxon's taphonomic signature include characteristics and locations of postmortem soft tissue or skeletal defects, the order in which different parts of the body are consumed or destroyed, and the timing and directionality of bodily disarticulation and scatter. Essentially, taphonomic signatures document the *general patterns* of damage that are associated with scavenging by a taxon, emerging from shared morphological, physiological, and behavioral characteristics. While this book emphasizes the variation in the signatures that arise from ecological influences on animal behavior, the rest of this chapter will break down how animal morphology, physiology, and instinctual behavior patterns are reflected in the scavenging evidence left at a death scene.

SOFT TISSUE AND BONE MODIFICATION

There are several morphological, physiological, and behavioral variables to consider when interpreting postmortem scavenging damage. Depending on the severity of the scavenging and the condition of the remains at the time of scavenging onset, such damage may be recorded in soft tissue, bone, or both. In addition to jaw shape (discussed earlier), dental morphology and habitual feeding behaviors of scavengers also play significant roles in shaping resultant damage.

Dental Morphology

Bite marks in soft tissue and tooth marks in bone naturally reflect the offending organism's dentition or beak. Consequently, marks produced by birds differ morphologically than those produced by mammals and aquatic organisms (i.e., fish, sharks, and some reptiles). Avian species do not have teeth, with damage inflicted instead by the hard tissue of the beak (Fig. 4.5). For example, in soft tissue, beaks inflict abrasions and punctures with hour-glass shapes corresponding to the upper and lower bills (Roll and Rous, 1991). In bone, avian damage is largely limited to surface scratches, although occasional punctures may be

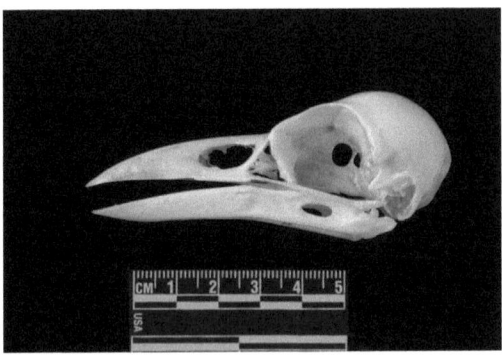

*Figure 4.5 American crow (*Corvus brachyrhynchos*) skull. Note the pointed beak.* Photograph by David Tuttle.

observed on elements where the cortical layer is thin, such as the delicate bones of the face (Reeves, 2009) (Fig. 5.16).

Aquatic scavengers, including sharks and crocodilians, are homodonts, meaning that all of their teeth have similar morphology. The teeth of most shark species are triangular, with many characterized by notched (saw-like) edges which produce wounds with serrated margins and "shaving" of underlying bone (Allaire et al., 2012; Ihama et al., 2009). In crocodilians, carnivorous reptiles including alligators and crocodiles, the teeth are more conical and when unworn, have distinctive vertical ridges in the enamel, called carinae, which produce unique bisected tooth marks in bone (Drumheller and Brochu, 2014; Njau and Blumenschine, 2006) (Fig. 4.6).

Conversely, heterodonts have different types of teeth. Terrestrial mammals have four types: incisors, canines, premolars, and molars, each with a distinctive form and function. Typically, incisors are broad and flat; canines are conical and specialized for puncturing; while premolars and molars have cusps which are adapted for shearing or grinding different types of food (i.e., meat, vegetation, or both). Although these tooth types are common among mammals, the precise form and the number of each type varies between taxa reflecting variations in diet composition. The number of each type present in the upper and lower jaws for each species are simplified into what is known as the dental formula. The dental formula divides the mouth into four

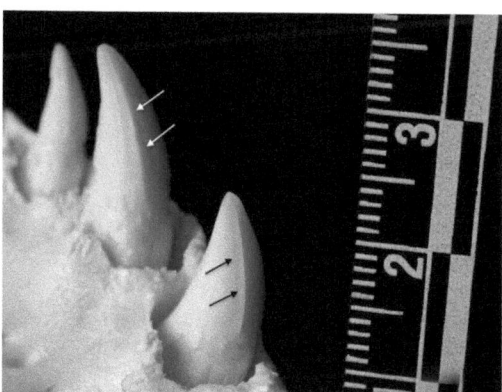

Figure 4.6 Carinae (vertical ridges—white and black *arrows) on maxillary teeth of an alligator (*Alligator mississipiensis*). Also note conical tooth morphology.* Photograph by David Tuttle.

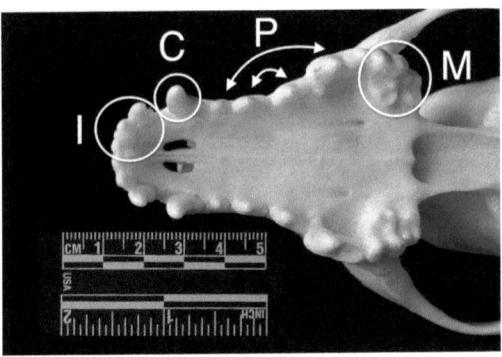

*Figure 4.7 Domestic dog (*Canis familiaris*) maxilla with tooth types indicated. I = incisors; C = canine; P = premolars; M = molars. The dental formula for a dog's upper dentition is 3I 1C 4P 2M. Also note the pointed and distinctive morphology of all the teeth.* Photograph by David Tuttle.

quadrants: upper right, upper left, lower right, and lower left. The number of incisors (I), canines (C), premolars (P), and molars (M) for one side (upper and lower) are then enumerated (Fig. 4.7). For example, the adult human dental formula is: $\frac{2}{2}I \; \frac{1}{1}C \; \frac{2}{2}P \; \frac{3}{3}M$, indicating that for each quadrant of the mouth, upper and lower, there are 2 incisors, 1 canine, 2 premolars, and 3 molars, for a total of 32 permanent teeth.

In addition, carnivorous mammals have a specialized adaptation of their last upper premolar and first lower molar, where the lower molar articulates to the inner edge of the upper premolar when the mouth is closed (Fig. 4.8). These teeth are known as carnassials, and give carnivores the ability to shear flesh via a scissoring action. The pointed cusps of carnassials are also responsible for leaving marks on bone, as carnivores use their back teeth for crushing and gnawing.

Generally, most terrestrial mammals leave four types of tooth marks on bone while scavenging: pits, punctures, scores, and furrows (Haglund et al., 1988; Pokines, 2014). Pits are indentations in the external layer of cortical bone created when pressure is applied from the tip or cusp of a tooth, with punctures produced when the cortical bone gives way (Fig. 5.3). Scores are scratches that occur when teeth drag across compact bone. When scores occur randomly across the diaphysis or bone shaft, they are termed gripping marks, produced when a carnivore readjusts the bone in their mouth for a better grip (Pokines and Symes, 2014) (Fig. 5.4). Furrows are deeper channels that are ultimately produced by repetitive gnawing, especially by canids and rodents. Furrows produced by these two taxa can be

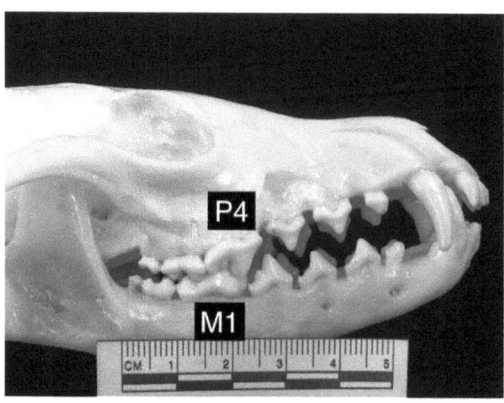

*Figure 4.8 Red fox (*Vulpes vulpes*) dentition with carnassial teeth indicated. Note how the lower first molar (M1) rests inside the fourth upper premolar (P4) when the mouth is closed.* Photograph by Elizabeth A. DiGangi.

distinguished by their shape and their association with other furrows, with paired furrows with flat bottoms indicating rodent involvement and irregular, V- or U-shaped furrows being characteristic of canids (Haglund, 1997a,b) (Figs. 5.8 [canids]; 5.47 [rodents]).

The relationship between the dimensions of individual tooth marks in bone (e.g., pit breadth or score length) and the dentition of the responsible species is obscured by overlaps in tooth size and shape within and between taxonomic groups (Delaney-Rivera et al., 2009). In addition, the size of a tooth mark is also affected by variables other than the size of the offending tooth, including bone size, density, and condition at the time of bite (i.e., green, dry, vs. weathered bone) or feeding behaviors of species within a taxon (Young et al., 2015). Consequently, the dimensions of individual tooth marks in bone cannot be used independently to identify a scavenger.

However, spatial patterning of tooth marks may assist in efforts to discriminate between scavenging taxa. For example, punctures in bone caused by bird beaks have a similar shape to those caused by the canine teeth of carnivorous mammals or the conical teeth of crocodilians. However, punctures caused by mammalian carnivores often occur in pairs with relative symmetry and predictable spatial relationships (i.e., the intercanine and interincisal distances discussed previously in this chapter), while those caused by birds are likely to occur in isolation (Komar and Beattie, 1998). Punctures created by crocodilians are relatively rare, and when they occur are largely limited to the epiphyseal regions of long bones, are often associated with compressed fractures, and may be bisected (Drumheller and Brochu, 2014).

Habitual Feeding Behaviors

In addition to dental morphology, the mechanics of carrion consumption may be reflected in bone modifications; however, research on how consumption patterns influence soft tissue injuries are lacking. Bites from species with stronger jaw musculature will be associated with more extensive crushing damage, and damaged areas are more likely to exhibit compression and spiral fractures in larger, stronger bones (Allaire et al., 2012; Carson et al., 2000; Harding and Wolf, 2006). Unique forms of mechanical food processing may also be evidenced in skeletal damage. For example, crocodilians are inertial feeders, using gravity to toss large portions of food to the back of their throats; the repetitious head movements required for inertial feeding produce diagnostic J-shaped "hook" scores on bone (Drumheller and Brochu, 2014). Deer also produce characteristic skeletal damage. The jaws of deer and other herbivorous ungulates, such as sheep, move transversely to efficiently shear vegetation. When chewing bone, this movement preferentially erodes the central aspect of the bone, creating a unique forked defect (Kierdorf, 1994) (Figs. 5.23 and 5.24).

Finally, unique feeding behaviors may create characteristic wound types, influencing how damage presents in soft tissue or bone. For example, the "death roll" of alligators allows the species to twist extremities off large prey by gripping and rolling along their own long axis; the result is amputation of the limb with shredded muscular, vascular, and nervous tissue (Drumheller and Brochu, 2014; Harding and Wolf, 2006). Some species of shark engage in a similar feeding behavior, biting and spinning along an appendage to shear away the soft tissue, which can leave spiraling gouges along the diaphyses of long bones (Ihama et al., 2009; Işcan and McCabe, 1995). A more familiar example, especially to dog owners, is the "kennel pattern" of damage, proposed to be motivated by boredom or stress. It is characterized by gnawing damage that is far more extensive than that observed in wild canid counterparts (Binford, 1981; Haynes, 1982; Rothschild and Schneider, 1997) (Fig. 5.8). Dental morphology and feeding behaviors are but two unique adaptations that contribute to unique soft tissue and skeletal defects in scavenged human remains; additional examples will be touched on in subsequent chapters.

PATTERNS OF TISSUE CONSUMPTION

Taphonomic signatures include patterns of soft tissue and bone destruction as well as the morphology of isolated defects. These patterns arise from variation in how different species preferentially utilize an available carcass, which reflects dietary requirements, digestive restrictions, and morphological or physiological characteristics of the scavenger that promote or prevent access to different tissues and organs. Nutritional quality of different body tissues vary, as they are composed of different proportions of carbohydrates, protein, and fat. The result is a hierarchy of tissue preference for scavenging species, theoretically determined both by the energetic quality of the tissue and the costs of obtaining and digesting it. Both energetic quality and cost, and thus tissue preference itself, are subject to ecological influences. Further, dietary requirements and the ways carrion is exploited to fulfill them may differ between populations due to ecological variables such as habitat type, climate, prey availability, and reproductive seasonality. Such deviations are discussed in Chapter 6, *Ecological Influences on Scavenging Behavior*. However, for the purposes of the chapter here, we briefly discuss how dietary requirements influence which carrion tissues are prioritized by scavenging species, as well as how access to some preferential tissues may be limited by species size and strength.

A consequence of evolutionary divergence of digestive capabilities is that different species have different nutritional requirements. As omnivores, brown bears (*Ursus arctos*) are adapted to diets with macronutrient compositions incorporating only 20% protein on average, with the remainder preferentially supplemented by fats or, when fats are unavailable, carbohydrates (Erlenbach et al., 2014). In contrast, more obligate carnivores (e.g., felids) may use protein to meet as much as 50% of their energy demands (Hewson-Hughes et al., 2011), while canids fall somewhere in the middle, with clear preferences for the animal protein that fulfills approximately 30% of their energetic needs (Anandarup et al., 2016; Hewson-Hughes et al., 2013). Given their macronutrient requirements, scavenging felids or canids are likely to ignore a carcass's skin in favor of protein-rich muscle and organ tissues

(Haglund, 1997a; Pickering, 2001). The larger and highly omnivorous bear species, however, are likely to prefer greater amounts of fatty tissue such as skin and subcutaneous fat (Bright, 2011; Nelson et al., 1983).

Other species engage in osteophagy, or bone eating, including the California condor (*Gymnogyps californianus*), gray squirrel (*Sciurus carolinensis*), desert tortoise (*Gopherus agassizii*), and deer (*Cervidae* spp.) (Cáceres et al., 2011; Collins et al., 2000; Klippel and Synstelien, 2007; Walde et al., 2007). These species consume dry bone to obtain important minerals such as calcium, phosphorous, and sodium; many of these species will ignore soft tissue altogether. Tooth marks left in bone by bone-eating species can be distinguished from carnivorous tooth marks by their association with cracking or microstriations and superimposition over weathered bone (Cáceres et al., 2011) (Fig. 5.47). As a final note, investigators should be aware that preferred tissues may not be accessible to all species. For example, with the exception of some vultures, many avian species are incapable of breaking the skin of large carrion to access preferred organ meat without the assistance of carnivorous competitors (Heinrich and Pepper, 1998; Stahler et al., 2002). Mesopredators or medium-sized predators such as the opossum are also largely limited to tissues accessible by way of natural orifices (i.e., the eyes, nose, and mouth) until the carcass is opened by the activity of invertebrates or other carnivores (King et al., 2016; Morton and Lord, 2006).

DISARTICULATION, SCATTERING, AND ELEMENT REMOVAL

Scavenging can also produce predictable patterns of disarticulation, scattering, and skeletal element survivorship. Elements that are disarticulated earlier during scavenging are more likely to be destroyed or carried away from a death scene, as separation from the remains promotes monopolized gnawing, transport, and caching (Haglund, 1997a). Skeletal elements that pass through a scavenger's digestive tract are likely to be damaged, if recovered at all (Montalvo et al., 2007; Pickering, 2001). However, skeletal element survivorship may also be biased by inadequate recovery efforts (Bright, 2011). For example, law enforcement officials may not always be trained in body recovery, and therefore may simply box up suspected human bones without completing a proper search of the area (i.e., one in line with archaeological techniques). If the remains have been scattered, it is likely that skeletal elements will go unnoticed. A biased recovery, either due to human inexperience or to scavenger interference, can adversely affect the overall investigation.

For species which swallow large portions of carrion, passage of skeletal elements through the digestive tract can decrease the integrity and identifiability of bone. The postdigestion preservation of bone and other tissues, such as hair or fingernails, depends on the digestive tract and strength of associated digestive enzymes. Tiger sharks (*Galeocerdo cuvier*), for example, may store food undigested in the stomach for up to 21 days (Rathbun and Rathbun, 1997). Remains can often be recovered directly from the stomach of tiger sharks fairly intact, although it is unclear how the digestive tract environment influences the rate of decomposition. In most scavenging species, however, exposure to gastric acids in the digestive tract results in corrosion, etching, and smoothing of skeletal elements (Pickering, 2001). The integrity (i.e., size and density) of ingested elements or fragments also impacts the probability of recovering skeletal material from scat and the ultimate preservation of digested bone. In mammalian carnivores, smaller elements are more likely to escape with only minimal mechanical damage (i.e., with that caused by chewing), while larger elements that are recovered will typically be severely damaged (Montalvo et al., 2007). For example, digits swallowed whole by large carnivores may be recovered intact, with the phalanges protected by tough layers of connective tissue (Murad and Boddy, 1987; Pickering and Carlson, 2004; Willey and Snyder, 1989).

Patterns of transport and the likelihood of caching body parts, skeletal elements, or personal effects are related to body size and strength. The smallest scavengers, for example mice, may not scatter or remove the remains at all. In contrast, large scavengers are able to move body parts or even entire bodies from their original deposition context. Sometimes the distances are considerable, but movement of the remains is not guaranteed. Groups of foraging vultures are also prone to scattering, with most elements recovered within 30 feet of the original deposition site, although a scatter area of 900 square feet is not uncommon (Spradley et al., 2012). While body size affects a scavenger's ability to move remains, predicting where skeletal elements may move to, how far from their original context they may be displaced, or whether or not they will be cached is highly dependent on local ecological variables and will be discussed further in Chapter 6.

CONCLUSION

Taphonomic signatures are predictable due to morphological, physiological, and behavioral traits that are shared among the species within taxonomic groups. Dentition and jaw morphology, dietary requirements, and habitual behaviors are ultimately reflected in the evidence of scavenger activity that is left at a death scene. The last of these, behavior, is highly subject to ecological influence and is discussed in greater detail later in this volume. In the next chapter, the taphonomic signatures of key members of the North American vertebrate scavenger guild are summarized, paired with brief discussions of possible variation produced under different ecological circumstances.

REFERENCES

Allaire, M.T., Manhein, M.H., Burgess, G.H., 2012. Shark-inflicted trauma: a case study of unidentified remains recovered from the Gulf of Mexico. J. Forensic Sci. 57 (6), 1675–1678.

Anandarup, B., Bhattacharjee, D., Paul, M., Singh, A., Gade, P.R., Shrestha, P., et al., 2016. The meat of the matter: a rule of thumb for scavenging dogs? Ethol. Ecol. Evol. 28 (4), 427–440.

Anderson, B., Manoukian, A., Holland, T., Grant, W., 2003. A death in paradise: human remains scavenged by a shark. In: Steadman, D. (Ed.), Hard Evidence: Case Studies in Forensic Anthropology, first ed. Pearson Education, Inc, Upper Saddle River, pp. 186–196.

Bernitz, H., Bernitz, Z., Steenkamp, G., Blumenthal, R., Stols, G., 2012. The individualisation of a dog bite mark: a case study highlighting the bite mark analysis, with emphasis on differences between dog and human bite marks. Int. J. Leg. Med. 126 (3), 441–446.

Binford, L.R., 1981. Patterns of bone modifications produced by nonhuman agents. In: Binford, L.R. (Ed.), Bones: Ancient Men and Modern Myths. Academic Press, San Diego, pp. 35–86.

Bowers, C.M., Johansen, R.J., 2002. Photographic evidence protocol: the use of digital imaging methods to rectify angular distortion and create life size reproductions of bite mark evidence. J. Forensic Sci. 47 (1), 178–185.

Bowers, C.M., Pretty, I.A., 2009. Expert disagreement in bitemark casework. J. Forensic Sci. 54 (4), 915–918.

Bright, L.N., 2011. Taphonomic signatures of animal scavenging in northern California: a forensic anthropological analysis. Unpublished Masters thesis. Department of Anthropology, California State University, Chico.

Bush, M.A., Thorsrud, K., Miller, R.G., Dorion, R.B., Bush, P.J., 2010. The response of skin to applied stress: investigation of bitemark distortion in a cadaver model. J. Forensic Sci. 55 (1), 71–76.

Cáceres, I., Esteban-Nadal, M., Bennàsar, M., Fernández-Jalvo, Y., 2011. Was it the deer or the fox? J. Archaeol. Sci. 38 (10), 2767–2774.

Carson, E.A., Stefan, V.H., Powell, J.F., 2000. Skeletal manifestations of bear scavenging. J. Forensic Sci. 45 (3), 515–526.

Clement, J., Blackwell, S., 2010. Is current bite mark analysis a misnomer? Forensic Sci. Int. 201 (1), 33–37.

Collins, P.W., Noel, F.R.S., Emslie, S.D., 2000. Faunal remains in California condor nest caves. The Condor 102 (1), 222–227.

Delaney-Rivera, C., Plummer, T.W., Hodgson, J.A., Forrest, F., Hertel, F., Oliver, J.S., 2009. Pits and pitfalls: taxonomic variability and patterning in tooth mark dimensions. J. Archaeol. Sci. 36 (11), 2597–2608.

Dove, C.J., Koch, S.L., 2011. Microscopy of feathers: a practical guide for forensic feather identification. Microscope-Chicago 59 (2), 51–71.

Drumheller, S.K., Brochu, C.A., 2014. A diagnosis of Alligator mississippiensis bite marks with comparisons to existing crocodylian datasets. Ichnos 21 (2), 131–146.

Elbroch, M., 2003. Mammal Tracks & Sign: A Guide to North American Species. Stackpole Books, Mechanicsburg.

Elbroch, M., Marks, E., 2001. Bird Tracks & Sign: A Guide to North American Species. Stackpole Books, Mechanicsburg.

Erlenbach, J.A., Rode, K.D., Raubenheimer, D., Robbins, C.T., 2014. Macronutrient optimization and energy maximization determine diets of brown bears. J. Mammal. 95 (1), 160–168.

Farley, S., Talbot, S.L., Sage, G.K., Sinnot, R., Coltrane, J., 2014. Use of DNA from bite marks to determine species and individual animals that attack humans. Wildl. Soc. Bull. 38 (2), 370–376.

Fonseca, G.M., Mora, E., Lucena, J., Cantin, M., 2015. Forensic studies of dog attacks on humans: a focus on bite mark analysis. Res. Rep. Forensic Med. Sci. 5, 39–51.

Foran, D.R., Crooks, K.R., Minta, S.C., 1997a. Species identification from scat: an unambiguous genetic method. Wildl. Soc. Bull. (1973-2006) 25 (4), 835–839.

Foran, D.R., Minta, S.C., Heinemeyer, K.S., 1997b. DNA-based analysis of hair to identify species and individuals for population research and monitoring. Wildl. Soc. Bull. 25 (4), 840–847.

Gilmour, R.J., Skinner, M.F., 2012. Forensic scatology: preliminary experimental study of the preparation and potential for identification of captive carnivore scat. J. Forensic Sci. 57 (1), 160–165.

Haglund, W.D., 1997a. Dogs and coyotes: postmortem involvement with human remains. In: Haglund, W.D., Sorg, M.H. (Eds.), Forensic Taphonomy: The Postmortem Fate of Human Remains. CRC Press, Boca Raton, pp. 367–382.

Haglund, W.D., 1997b. Rodents and human remains. In: Haglund, W.D., Sorg, M.H. (Eds.), Forensic Taphonomy: The Postmortem Fate of Human Remains. CRC Press, Boca Raton, pp. 405–414.

Haglund, W.D., Reay, D.T., Swindler, D.R., 1988. Tooth mark artifacts and survival of bones in animal scavenged human skeletons. J. Forensic Sci. 33 (4), 985–997.

Halfpenny, J., 2015a. Scats and Tracks of the Desert Southwest: A Field Guide to the Signs of 70 Wildlife Species, second ed. Rowman & Littlefield, FalconGuides.

Halfpenny, J., 2015b. Scats and Tracks of the Rocky Mountains: A Field Guide to the Signs of 70 Wildlife Species, third ed. Rowman & Littlefield, FalconGuides.

Harding, B.E., Wolf, B.C., 2006. Alligator attacks in southwest Florida. J. Forensic Sci. 51 (3), 674–677.

Haynes, G., 1982. Utilization and skeletal disturbances of North American prey carcasses. Arctic 35 (2), 266–281.

Heinrich, B., Pepper, J.W., 1998. Influence of competitors on caching behaviour in the Common Raven, Corvus corax. Anim. Behav. 56 (5), 1083−1090.

Hewson-Hughes, A.K., Hewson-Hughes, V.L., Colyer, A., Miller, A.T., McGrane, S.J., Hall, S. R., et al., 2013. Geometric analysis of macronutrient selection in breeds of the domestic dog, Canis lupus familiaris. Behav. Ecol. 24 (1), 293−304.

Hewson-Hughes, A.K., Hewson-Hughes, V.L., Miller, A.T., Hall, S.R., Simpson, S.J., Raubenheimer, D., 2011. Geometric analysis of macronutrient selection in the adult domestic cat, Felis catus. J. Exp. Biol. 214 (6), 1039−1051.

Hogan, F.E., Cooke, R., Burridge, C.P., Norman, J.A., 2008. Optimizing the use of shed feathers for genetic analysis. Mol. Ecol. Resour. 8 (3), 561−567.

Ihama, Y., Ninomiya, K., Noguchi, M., Fuke, C., Miyazaki, T., 2009. Characteristic features of injuries due to shark attacks: a review of 12 cases. Leg. Med. 11 (5), 219−225.

Işcan, M.Y., McCabe, B.Q., 1995. Analysis of human remains recovered from a shark. Forensic Sci. Int. 72 (1), 15−23.

Kierdorf, U., 1994. A further example of long-bone damage due to chewing by deer. Int. J. Osteoarchaeol. 4 (3), 209−213.

King, K.A., Lord, W.D., Ketchum, H.R., O'Brien, R.C., 2016. Postmortem scavenging by the Virginia opossum (Didelphis virginiana): impact on taphonomic assemblages and progression. Forensic Sci. Int. 266 (576), e1-576.e6.

Klippel, W.E., Synstelien, J.A., 2007. Rodents as taphonomic agents: bone gnawing by brown rats and gray squirrels. J. Forensic Sci. 52 (4), 765−773.

Komar, D., Beattie, O., 1998. Identifying bird scavenging in fleshed and dry remains. J. Canadian Soc. Forensic Sci. 31 (3), 177−188.

Lowry, D., de Castro, A.L.F., Mara, K., Whitenack, L.B., Delius, B., Burgess, G.H., et al., 2009. Determining shark size from forensic analysis of bite damage. Mar. Biol. 156, 2483.

Marden, K., Sorg, M., Haglund, W., 2013. Taphonomy. In: DiGangi, E., Moore, M. (Eds.), Research Methods in Human Skeletal Biology. Academic Press, San Diego, pp. 241−262.

Martin-de las Heras, S., Valenzuela, A., Ogayar, C., Valverde, A.J., Torres, J.C., 2005. Computer-based production of comparison overlays from 3D-scanned dental casts for bite mark analysis. J. Forensic Sci. 50 (1), 226−227.

Melton, T., Holland, C., 2007. Routine forensic use of the mitochondrial 12S ribosomal RNA gene for species identification. J. Forensic Sci. 52 (6), 1305−1307.

Montalvo, C.I., Pessino, M.E.M., González, V.H., 2007. Taphonomic analysis of remains of mammals eaten by pumas (Puma concolor Carnivora, Felidae) in Central Argentina. J. Archaeol. Sci. 34 (12), 2151−2160.

Morton, R.J., Lord, W.D., 2006. Taphonomy of child-sized remains: a study of scattering and scavenging in Virginia, USA. J. Forensic Sci. 51 (3), 475−479.

Murad, T., 1997. The utilization of faunal evidence in the recovery of human remains. In: Haglund, W.D., Sorg, M.H. (Eds.), Forensic Taphonomy: The Postmortem Fate of Human Remains. CRC Press, Boca Raton, pp. 395−404.

Murad, T.A., Boddy, M.A., 1987. A case with bear facts. J. Forensic Sci. 32 (6), 1819−1826.

Murie, O.J., Elbroch, M., 2005. The Peterson Field Guide to Animal Tracks, third ed. Houghton Mifflin Harcourt, New York.

Murmann, D.C., Brumit, P.C., Schrader, B.A., Senn, D.R., 2006. A comparison of animal jaws and bite mark patterns. J. Forensic Sci. 51 (4), 846−860.

Nelson, R.A., Folk, G.E., Pfeiffer, E.W., Craighead, J.J., Jonkel, C.J., Steiger, D.L., 1983. Behavior, biochemistry, and hibernation in black, grizzly, and polar bears. Bears: Their Biology and Management 5, 284–290.

Njau, J.K., Blumenschine, R.J., 2006. A diagnosis of crocodile feeding traces on larger mammal bone, with fossil examples from the Plio-Pleistocene Olduvai Basin, Tanzania. J. Hum. Evol. 50 (2), 142–162.

Page, M., Taylor, J., Blenkin, M., 2012. Context effects and observer bias—implications for forensic odontology. J. Forensic Sci. 57 (1), 108–112.

Pickering, T.R., 2001. Carnivore voiding: a taphonomic process with the potential for the deposition of forensic evidence. J. Forensic Sci. 46 (2), 406–411.

Pickering, T.R., Carlson, K.J., 2004. Baboon taphonomy and its relevance to the investigation of large felid involvement in human forensic cases. Forensic Sci. Int. 144 (1), 37–44.

Pokines, J., 2014. Faunal dispersal, reconcentration, and gnawing damage to bone in terrestrial environments. In: Pokines, J., Symes, S. (Eds.), Manual of Forensic Taphonomy. CRC Press, Boca Raton, pp. 201–248.

Pokines, J., Symes, S.A., 2014. (Eds.), Manual of Forensic Taphonomy. CRC Press, Boca Raton.

Pretty, I., Sweet, D., 2001. The scientific basis for human bitemark analyses—a critical review. Sci. Just. 41 (2), 85–92.

Pretty, I.A., 2006. The barriers to achieving an evidence base for bitemark analysis. Forensic Sci. Int. 159, S110–S120.

Rathbun, T., Rathbun, B., 1997. Human remains recovered from a shark's stomach in South Carolina. In: Haglund, W.D., Sorg, M.H. (Eds.), Forensic Taphonomy: The Postmortem Fate of Human Remains. CRC Press, Boca Raton, pp. 249–256.

Reeves, N.M., 2009. Taphonomic effects of vulture scavenging. J. Forensic Sci. 54 (3), 523–528.

Roll, P., Rous, F., 1991. Injuries by chicken bills: characteristic wound morphology. Forensic Sci. Int. 52 (1), 25–30.

Rollins, C., Spencer, D., 1995. A fatality and the American mountain lion: bite mark analysis and profile of the offending lion. J. Forensic Sci. 40 (3), 486–489.

Rothschild, M.A., Schneider, V., 1997. On the temporal onset of postmortem animal scavenging:"Motivation" of the animal. Forensic Sci. Int. 89 (1), 57–64.

Rothwell, B.R., Thien, A., 2001. Analysis of distortion in preserved bite mark skin. J. Forensic Sci. 46 (3), 573–576.

Rudnick, J.A., Katzner, T.E., Bragin, E.A., DeWoody, J.A., 2007. Species identification of birds through genetic analysis of naturally shed feathers. Mol. Ecol. Resour. 7 (5), 757–762.

Santoro, V., Lozito, P., De Donno, A., Introna, F., 2011a. Experimental study of bite mark injuries by digital analysis. J. Forensic Sci. 56 (1), 224–228.

Santoro, V., Smaldone, G., Lozito, P., Smaldone, M., Introna, F., 2011b. A forensic approach to fatal dog attacks. A case study and review of the literature. Forensic Sci. Int. 206 (1), e37–e42.

Sato, I., Nakaki, S., Murata, K., Takeshita, H., Mukai, T., 2010. Forensic hair analysis to identify animal species on a case of pet animal abuse. Int. J. Leg. Med. 124 (3), 249–256.

Schulz, I., Schneider, P.M., Olek, K., Rothschild, M.A., Tsokos, M., 2006. Examination of postmortem animal interference to human remains using cross-species multiplex PCR. Forensic Sci. Med. Pathol. 2 (2), 95–101.

Sessions, B.D., Hess, W.M., Skidmore, W.S., 2009. Can hair width and scale pattern and direction of dorsal scapular mammalian hair be a relatively simple means to identify species? J. Nat. Hist. 43 (7-8), 489–507.

Sheasby, D., MacDonald, D., 2001. A forensic classification of distortion in human bite marks. Forensic Sci. Int. 122 (1), 75–78.

Souviron, R.R., 2011. Animal bites. In: Dorion, R.B.J. (Ed.), Bitemark Evidence, second ed. CRC Press, Boca Raton, pp. 209–216.

Spradley, M.K., Hamilton, M.D., Giordano, A., 2012. Spatial patterning of vulture scavenged human remains. Forensic Sci. Int. 219, 57–63.

Stahler, D., Heinrich, B., Smith, D., 2002. Common ravens, Corvus corax, preferentially associate with grey wolves, Canis lupus, as a foraging strategy in winter. Anim. Behav. 64 (2), 283–290.

Thali, M., Braun, M., Markwalder, T.H., Brueschweiler, W., Zollinger, U., Malik, N.J., et al., 2003. Bite mark documentation and analysis: the forensic 3D/CAD supported photogrammetry approach. Forensic Sci. Int. 135 (2), 115–121.

Tsai, L., Huang, M., Hsiao, C., Lin, C.A., Chen, S., Lee, J.C., et al., 2007. Species identification of animal specimens by cytochrome b gene. Forensic Sci. J. 6 (1), 63–66.

Walde, A.D., Delaney, D.K., Harless, M.L., Pater, L.L., 2007. Osteophagy by the desert tortoise (Gopherus agassizii). Southw. Natural. 52 (1), 147–149.

Willey, P., Snyder, L., 1989. Canid modification of human remains: implications for time-since-death estimations. J.Forensic Sci. 34 (4), 894–901.

Young, A., Stillman, R., Smith, M.J., Korstjens, A.H., 2015. Scavenger species-typical alteration to bone: using bite mark dimensions to identify scavengers. J. Forensic Sci. 60 (6), 1426–1435.

CHAPTER 5

What Big Teeth You Have: Taphonomic Signatures of North American Scavengers

There are several taxa from vertebrate species of the North American scavenger guild that are important from a forensic standpoint. Here, we present a summarized account of their habitats, distribution, behavior, morphology, and taphonomic signatures. Taxa are ordered alphabetically by order, family, or species name.

FAMILY CANIDAE—CANIDS—DOGS, COYOTES, FOXES, AND WOLVES

Canid species scavenge for nutrition, with a preference for fresh carrion free of bloat, insect activity, or adipocere (Haglund, 1997a). The probability of scavenging by different canid species varies, with

Forensic Taphonomy and Ecology of North American Scavengers. DOI: https://doi.org/10.1016/B978-0-12-813243-2.00005-1

upper-level predators (e.g., gray wolves, *Canis lupus*) (Figs. 5.1−5.2) depending heavily on hunted prey and scavenging infrequently while domestic dogs (*Canis familiaris*), when uncared for, may subsist almost entirely on scavenged food sources, including human refuse (Anandarup et al., 2016; Metz et al., 2012). Coyotes (*Canis latrans*,

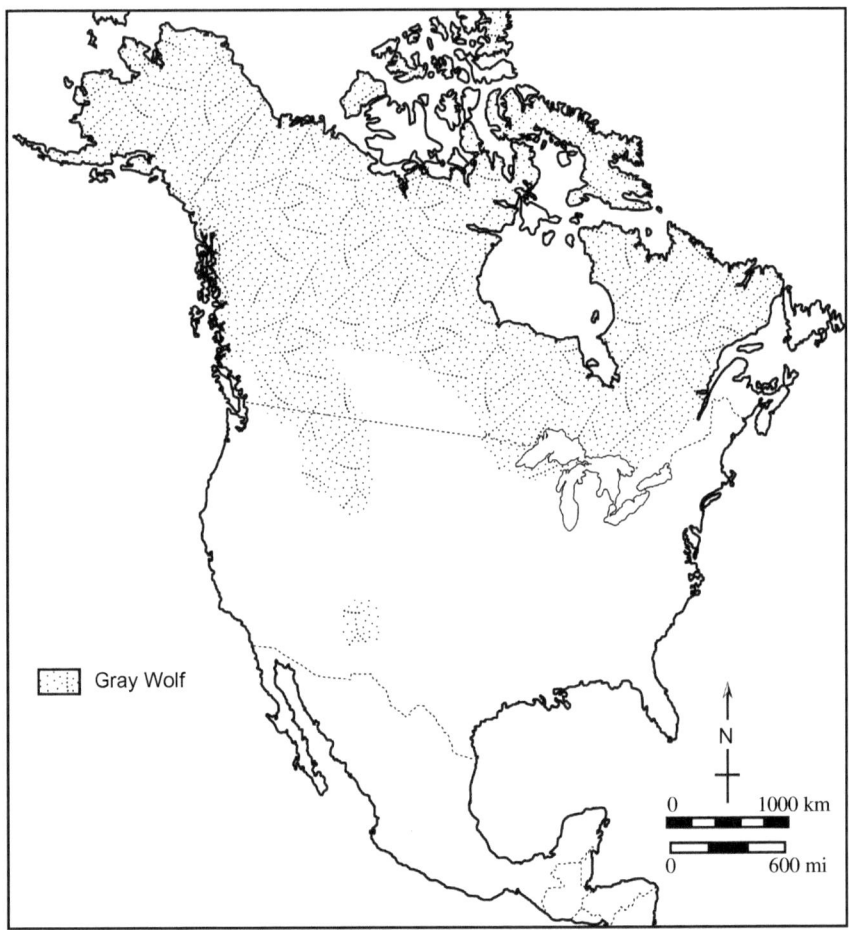

Figure 5.1 Distribution of the gray wolf (C. lupus) in North America. Note that several depicted areas are regions where the wolf has been reintroduced to parts of its native range (the region in Arizona/New Mexico contains reintroduced Mexican gray wolves (C. lupus baileyi), an endangered subspecies of gray wolves). Not depicted are coyotes, various species of fox, and domestic dogs, all of which are found ubiquitously throughout the continent. Distribution maps are meant only for general informational use, as species distributions change over time and individuals may occasionally be found outside of the depicted boundaries. Distribution information compiled from Reid, F.A., 2006. The Princeton Field Guide to Mammals of North America. Houghton Mifflin Harcourt, Boston; IUCN 2017. The IUCN Red List of Threatened Species. Version 2017-1. Available from: http://www.iucnredlist.org/ (accessed 31.07.17) (IUCN, 2017).

Table 5.1 Common North American Canid Species

Species Name	Conservation Status	Habitat
Red fox (*V. vulpes*)	Least concern	Prefers mixed habitats of brush and fields, highly adaptable to alternative environments.
Coyote (*C. latrans*)	Least concern	Highly adaptable and widespread, most abundant in mixed habitats.
Gray wolf (*C. lupus*)	Least concern	Heavy forest and tundra.
Domestic dog (*C. familiaris*)		Various breeds found as pets throughout Canada, the United States, and Mexico.

Conservation status courtesy of IUCN Red List of Threatened Species.
Habitat information adapted from Reid, F.A., 2006. The Princeton Field Guide to Mammals of North America. Houghton Mifflin Harcourt, Boston.

Table 5.2 Other North American Canid Species

Species Name	Distribution
Arctic fox (*V. lagopus*)	Circumpolar arctic (Northern Canada and Alaska)
Gray fox (*U. cinereoargenteus*)	Western, central, and eastern United States through Mexico
Kit fox (*V. macrotis*)	Southwestern United States and northern Mexico
Swift fox (*V. velox*)	Small populations along the western United States −Canada border, Wyoming and south through northern Texas (United States)
Eastern timber wolf (*C. lycaon*)	Southeastern Canada, northeastern United States

Distribution information adapted from Reid, F.A., 2006. The Princeton Field Guide to Mammals of North America. Houghton Mifflin Harcourt, Boston.

Table 5.3 Behavioral and Morphological Characteristics of Common Canids

	Red Fox	Coyote	Gray Wolf
Peak daily activity[a]	Nocturnal or crepuscular	Mixed	Crepuscular
Sociality[a]	Solitary foragers	Up to 7 in stable packs	Up to 15 in stable packs
Fur[a]	Orange-red	Gray in north, tawny in south	Gray, with variations
Weight (lbs) [a]	8−15	20−50	60−154
Dental formula[b]	$\frac{3}{3}I\frac{1}{1}C\frac{4}{4}P\frac{2}{3}M$		

[a]*Reid (2006).*
[b]*Adams and Crabtree (2011).*

*Figure 5.2 Gray wolf (*Canis lupus*).* Photograph by R.T. Kramer.

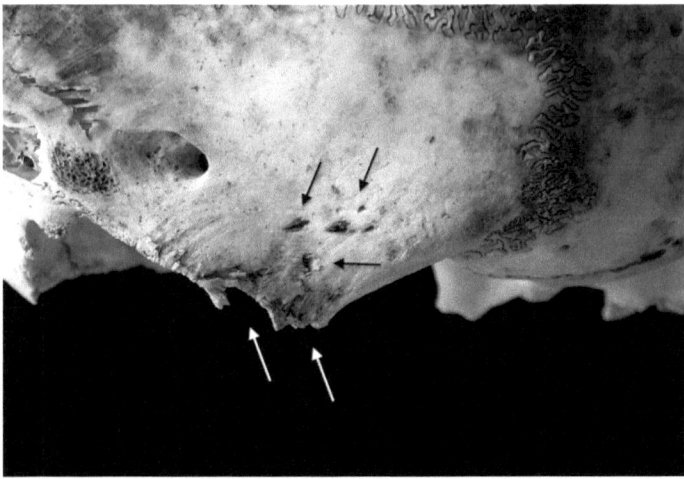

*Figure 5.3 Deer (Cervidae spp.) cranium exhibiting punctures (*black *arrows) on the left frontal bone, with gnawing along the upper edge of the eye orbit (*white *arrows), both likely due to a small canid. Punctures range from 2 to 3 mm in width. Caudal is to the right.* Photograph by David Tuttle.

Fig. 5.5) and various species of fox (e.g., *Urocyon cinereoargenteus*, Fig. 5.6) often scavenge at an intermediate level.

Like other carnivores, canid-scavenged skeletal remains are marked by pits, punctures, scores, and furrows (Colard et al., 2015; Haglund, 1997a; Steadman and Worne, 2007) (Figs. 5.3 and 5.4). In long bones,

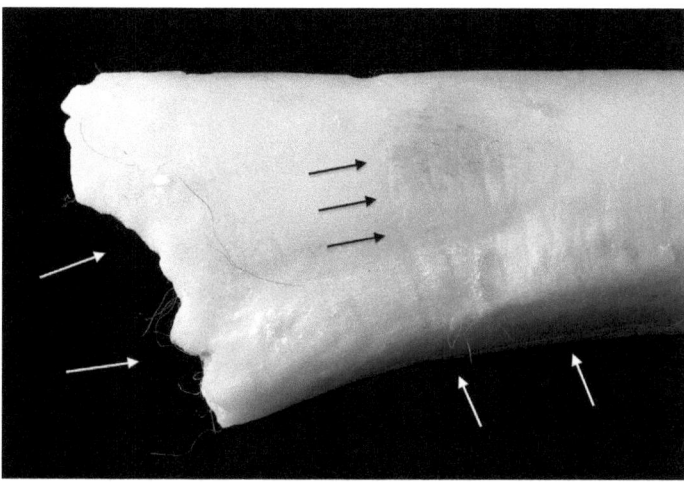

Figure 5.4 Nonhuman mammal bone gnawed by large (c. 55–80 lbs.) domestic dogs. Note irregular end and adherent fur (left arrows), shallow gripping marks along shaft (black arrows), and deeper scoring marks towards bottom edge (bottom arrows). The bone fragment is approximately 9 × 5 cm. Photograph by David Tuttle.

Figure 5.5 Coyote (Canis latrans). "Coyote" is in the public domain.

most tooth marks are concentrated at epiphyses, although scoring is more common on the cortical bone of the shaft (Colard et al., 2015; Young et al., 2015). With advanced gnawing, the trabecular bone of the epiphyses may be "scooped out," leaving only the cylindrical diaphysis with smooth edges and extensive furrowing into the marrow

*Figure 5.6 Gray fox (*Urocyon cinereoargenteus*). Photograph by David Tuttle.*

cavity (Haglund, 1997a; Pokines, 2014; Willey and Snyder, 1989). Destruction of smaller bony features, such as the processes of the vertebrae, is also common, and stronger breeds may also fracture flat bones (e.g., the scapula) or smaller long bones (e.g., the radius, ulna, or fibula) (Haglund, 1997a; Young et al., 2015). See Tables 5.1–5.3 for habitats, distributions, and behavioral and morphological characteristics of common North American canid species. A summary of bone modifications characteristic of scavenging by canids can be found in Table 5.4.

In soft tissue, the margins of wounds typically have borders that are irregular and notched or wrinkled (Colard et al., 2015; Haglund, 1997a,b; Rothschild and Schneider, 1997; Tsokos and Schulz, 1999). The irregularity of soft tissue wounds is the result of a hole-and-tear feeding pattern, in which the prominent canines are anchored into the tissue before shaking the head to tear it from the body (Santoro et al., 2011). The punctures produced by these canines are typically V-shaped in cross section and are often accompanied by stretch lacerations (Santoro et al., 2011; Tsokos and Schulz, 1999). Exposed muscle tissue may be "polished" with a wet sheen created by repetitive licking, particularly in cases of indoor scavenging by domestic dogs (Colard et al., 2015). See Table 5.5 for a summary of characteristics of canid scavenging in soft tissue.

Table 5.4 Bone Modification by Canids

Damage Category	Traits	Common Locations
Margins of damage	Crushed bone[a]	Scapular borders
		Sacral margins
Tooth marks	Pits[a,b,c]	Generalized: cortical, and trabecular bone
	Punctures[a,b,c]	Facial bones
		Long bone epiphyses
		Thin, flat bones
	Scores[a,c,d]	Long bone diaphyses
	Furrowing[a,b,c,d]	Long bone epiphyses
		Marrow cavities
Breakage	"Scooping" of trabecular bone[a,e]	Long bone epiphyses
	Obliteration of features[a,c]	Vertebral processes
		Mandibular processes
		Proximal rib ends
	Depressed fractures[a]	Flat bones
	Other fractures[a,c]	Diaphyses of smaller long bones (e.g., radius, ulna, fibula)

[a]Haglund (1997a).
[b]Steadman and Worne (2007).
[c]Young et al. (2015).
[d]Colard et al. (2015).
[e]Willey and Snyder (1989).

Table 5.5 Soft Tissue Defects Produced by Canids

Damage Category	Traits
Margins	Irregular, wrinkled, or notched borders[a,b,c,d]
	Wound "polishing" by repeated licking[d]
Claw and tooth marks	Superficial scratches beyond defect borders[c]
	V-shaped punctures[c,d]
Tissue injury	Stretch lacerations[e]

[a]Haglund (1997a).
[b]Rothschild and Schneider (1997).
[c]Tsokos and Schulz (1999).
[d]Colard et al. (2015).
[e]Santoro et al. (2011).

The pattern of bodily consumption for canids has been well documented. In canid scavenging of human remains, soft tissue of the face and neck region is consumed first, followed by the thoracic organs and disarticulation of the upper and lower limbs

(Haglund, 1997a,b; Willey and Snyder, 1989). This is distinct from canid scavenging of ungulate carcasses, where the thoracic cavity is targeted first and the soft tissues of the head and neck are among the last to be consumed; this may be a result of human utilization of clothing, which could impede scavenger access to the preferred thoracic tissues (Haynes, 1982; Pokines, 2014). In an experimental study of the scavenging of monitored deer carcasses, red foxes (*Vulpes vulpes*) were observed to have a notably different consumption pattern, focusing on the hindlimbs first, followed by the forelimbs, before scavenging the thoracic cavity (Young et al., 2015). One exception to this general pattern occurred in a carcass that was first scavenged by a domestic dog, creating an opening in the thoracic cavity which was subsequently used by scavenging foxes (Young et al., 2015). This may indicate that foxes are unable to open the thoracic cavity without assistance from larger carnivores, which may alter their consumption pattern of human remains.

It is common for canids to scatter or remove skeletal elements from their original location. Elements are often scattered along game trails, moving away from human activity and toward areas of visual cover or dens (Kjorlien et al., 2009; Pokines, 2014; Young et al., 2015). Foxes will attempt to move carcasses by dragging them by the appendages, and are known for their tendency to re-scavenge skeletonized remains, with distance and degree of scatter increasing after each visit (Young et al., 2015). Some canid species, including wolves, will remove and transport portions of a carcass to feed pregnant mothers or pups, and thus it is possible that skeletal elements or body parts may be recovered from wolf dens during whelping season (Bankaitis, 2012). Canids will also cache food, burying body parts or entire carcasses under mounds of soil or snow, which may be marked with leaf litter or twigs (Nelson, 2011; Vander Wall, 1990; Young et al., 2015).

Domestic Dogs and Indoor Scavenging

Contrary to outdoor-scavenged remains, cases of indoor scavenging by domestic dogs (Fig. 5.7) follows a unique pattern, with damage often limited to the head, neck, and upper extremities. A review of 41 forensic cases involving scavenging by domestic dogs identified these patterned differences in behavior between outdoor and indoor contexts (Colard et al., 2015). This distinction has been attributed to motivational differences between the scavenging events, first suggested in the

*Figure 5.7 Domestic dogs (*Canis familiaris*).* Photograph by Elizabeth A. DiGangi.

literature by Rothschild and Schneider (1997). These authors reported a forensic case in which a man died with his dog locked in the room. The animal subsequently scavenged areas of the owner's face and neck, despite having access to its normal food supply. The fact that the dog's vomit later indicated that his stomach contained dog food, and an esti- mated postmortem interval (PMI) of only 45 minutes, lent additional credence to the hypothesis that the scavenging was motivated by some- thing other than hunger (Rothschild and Schneider, 1997). Almost 20 years later, in their retrospective review of forensic casework involving scavenging by pet dogs, Colard et al. (2015) also favored this interpreta- tion and offered psychological distress in response to the death of an owner as a possible alternative motive for scavenging by dogs. In this interpretation, the animal becomes distressed at the person's unrespon- siveness and begins licking the face in an attempt to arouse them. When this does not work, the distress increases and the licking turns into biting and chewing.

Domestic dogs are also known to continue gnawing skeletal ele- ments long after they have been depleted of nutrients. In the wild, skel- etal damage inflicted by canids is often concentrated at long bone epiphyses, a consequence of attempts to extract nutritious grease from trabecular bone (Binford, 1981; Haglund, 1997a). Conversely, gnawing by domestic dogs (i.e., the "kennel pattern") is hypothesized to be

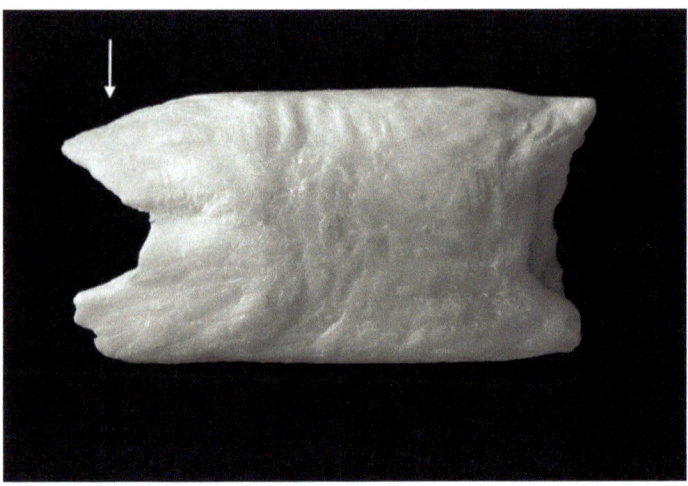

Figure 5.8 Nonhuman mammal bone gnawed by large (c. 55–80 lbs.) domestic dogs, exhibiting the common dog pattern of kennel damage as well as characteristic canid marks. Note the "penciling" of the shaft (arrow), numerous scoring marks and deeper furrows dispersed across the shaft, and irregular ends. The bone fragment is approximately 5 × 9 cm. Photograph by David Tuttle.

motivated by boredom or stress, rather than grease extraction. Therefore skeletal damage produced by domestic dogs is often far more extensive, with bones reduced to shafts or splinters and edges polished from repeated gnawing (Haynes, 1982; Pokines, 2014) (Figs. 5.8 and 5.9). In addition, "penciling" of the shaft, where one end of the diaphyses is narrowed due to wearing of cortical bone, can occur. An experimental study by Binford (1981) demonstrates the kennel pattern, as bones collected from wolf dens fared better than those introduced to dog yards.

A case reported by Steadman and Worne (2007) is an excellent example of the substantial destruction that gnawing by domestic dogs may produce in the forensic context. In this case, human remains were scavenged by the decedent's two pet dogs. Within a month of death, the dogs had consumed almost all soft tissue and reduced the skeletal remains to the calvarium, the shafts of a few long bones, and bone splinters. The absence of decomposition odors or stains present at the death scene indicates that the onset of scavenging by the dogs was particularly rapid in this case, occurring during the fresh stage of decomposition. This evidence also supports the onset of distress scavenging hypothesis, discussed earlier.

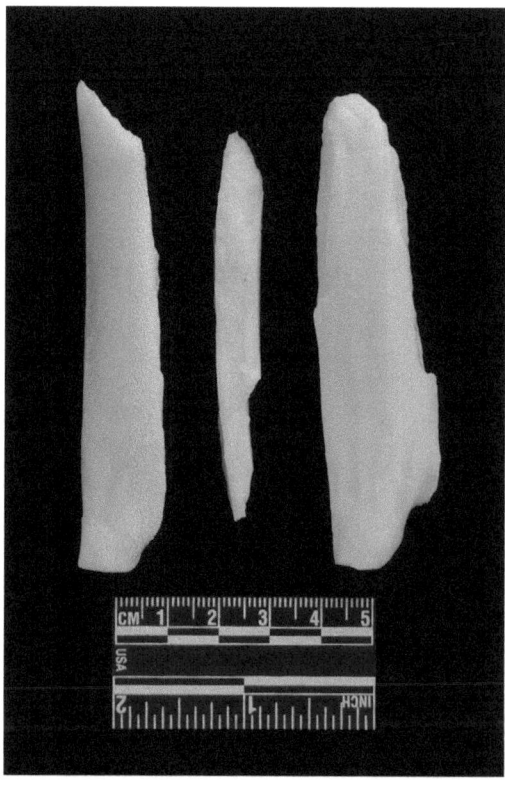

Figure 5.9 Bone splinters resulting from the destruction of a nonhuman mammal bone gnawed by large (c. 55–80 lbs.) domestic dogs. Photograph by David Tuttle.

Distinguishing Between Dog Attacks and Scavenging

It is important to distinguish between perimortem attacks (i.e., around the time of death) by canids and postmortem (after death) scavenging, especially given the frequency of human interactions with domestic dogs. If dog scavenging is mistaken for a dog attack, the suspect animal or animals may be wrongfully classified as a threat to public safety and euthanized (Hernández-Carrasco et al., 2016). Dog attack indicators differ from postmortem scavenging in several respects. Bite marks produced in dog attacks may be superimposed, as dogs may readjust their bite if their grip is lost while a victim attempts to escape (Fonseca et al., 2015). Defensive injuries will often be found on the forearms of the victim of a dog attack and dogs will often target the lower limbs in an attempt to bring down an individual, followed by concentrated attacks on the head and neck (Fonseca et al., 2015; Fonseca and Palacios, 2013). As dogs will typically scavenge the appendages last, the presence of such injuries without significant damage to the face, neck, or torso can serve as a primary indication that damage was inflicted perimortem. Such primary evidence can be followed up with analysis of vital responses in tissue excised from wounds, as attacking dogs will often begin consuming a victim before they are dead (Fonseca et al., 2015). In cases where a dog attack is suspected, investigators may find it useful to contact a forensic veterinarian and/or animal behavior specialist, as suggested by Hernández-Carrasco et al. (2016).

Canid Summary

- Distribution of gray wolves is across Canada and Alaska, in the Northern Rockies, Great Lakes, and Pacific Northwest, with much smaller populations of reintroduced Mexican gray wolves and red wolves in different parts of the United States
- Coyotes, foxes, and domestic dogs are ubiquitous throughout the continent
- Canids are preferentially carnivorous, although domestic dogs can survive on an omnivorous diet
- Wolves prefer hunting and rarely scavenge, with other canids scavenging more frequently
- Wolves, coyotes, and dogs (in outdoor contexts) have a distinct disarticulation sequence from that of foxes
- Dogs kept indoors exhibit a distinct scavenging pattern, important to distinguish from the pattern left by a dog attack
- Scattering and caching is common by all canid species
- Bone damage includes pits, punctures, scores, and furrows; as well as edge polish and bone splinters, particularly with domestic dogs. Such damage tends to cluster at epiphyseal ends, although there are exceptions. Destruction of small bones or features is also common.

FAMILY CATHARTIDAE—NEW WORLD VULTURES

As discussed elsewhere in this volume, vultures are the only extant families that are obligate scavengers, or animals whose sole feeding strategy is the consumption of carrion. There are three species of North American vultures, and there are distinct behavioral differences between them related to how they obtain carrion (Tables 5.6 and 5.7). See Fig. 5.10 for the distributions of New World vulture

Table 5.6 North American Vulture Species

Species Name	Conservation Status	Habitat
American black vulture (*C. atratus*)	Least concern	Mixed forests with open areas.
Turkey vulture (*C. aura*)	Least concern	Fields, open countryside.
California condor (*G. califorianus*)	Critically endangered	Pacific beaches, mountain forests, and meadows.

Data courtesy of Cornell Lab of Ornithology. 2015. All About Birds. Cornell University. Available from: https://www.allaboutbirds.org/ (accessed 31.07.17).

Table 5.7 Behavioral and Morphological Characteristics of Vultures

	American Black Vulture	Turkey Vulture	California Condor
Peak daily activity	Diurnal		
Sociality	Roost, forage in large groups	Roost in large groups, solitary foragers	Roost, forage in large groups
Plumage	Black	Dark brown	Black
Wingspan (cm)	137–150	170–178	Approximately 280
Weight range (lbs)	3.5–4.9	Approximately 4.4	15.4–21.8
Notable characteristics	Black, featherless head	Red, featherless head	Yellow-orange plumage on head
	Lateral, inferior wingtips are white	Pale inferior flight feathers	

Data courtesy of Cornell Lab of Ornithology. 2015. All About Birds. Cornell University. Available from: https://www.allaboutbirds.org/ (accessed 31.07.17).

species in North America. American black vultures (*Coragyps atratus*) (Fig. 5.11) demonstrate a preference for large carrion (Coleman and Fraser, 1987). This preference is supported by group foraging and their high tolerance for conspecifics, as larger carcasses can easily feed multiple birds (Buckley, 1996). In contrast, turkey vultures (*Cathartes aura*) (Fig. 5.12) have heightened olfactory sensitivity that allow them to detect, locate, and begin feeding on carrion more quickly, therefore improving their ability to locate smaller carcasses (Coleman and Fraser, 1987; Kiff, 2000). Because turkey vultures are adapted to exploit smaller carcasses that often cannot be shared between birds, they typically forage alone or in small groups (Prior and Weatherhead, 1991) (Fig. 5.13). The enhanced olfactory senses of turkey vultures also promote location of carrion hidden by leaf litter or within shallow burials, which vultures can access by scraping away debris with their feet (Klein, 2013; Platt et al., 2015; Smith et al., 2002).

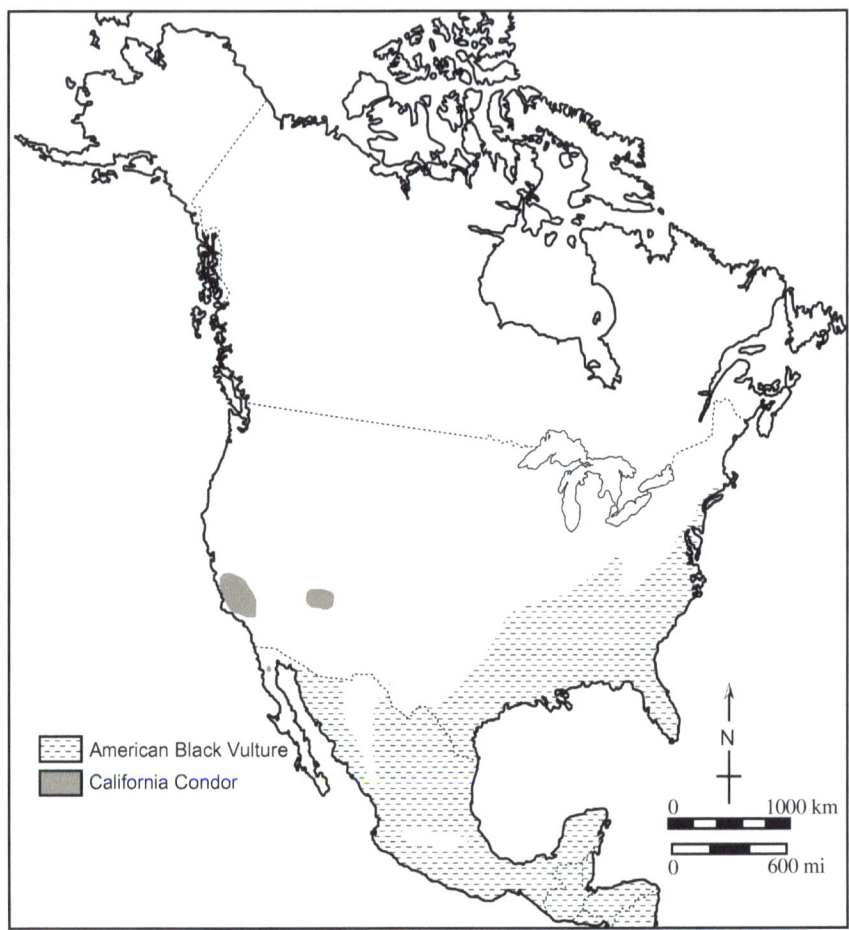

Figure 5.10 Distribution of the American black vulture (C. atratus) and California condor (G. califorianus) in North America. Turkey vultures (C. aura), not depicted, are widespread throughout the southern part of the continent and migrate as far north as southern Canada during the summer. Distribution maps are meant only for general informational use, as species distributions change over time and individuals may occasionally be found outside of the depicted boundaries. Distribution information compiled from Cornell Lab of Ornithology. 2015. All About Birds. Cornell University. Available from: https://www.allaboutbirds.org/ (accessed 31.07.17); IUCN 2017. The IUCN Red List of Threatened Species. Version 2017-1. Available from: http://www.iucnredlist.org/ (accessed 31.07.17).

The largest bird in North America, the California condor (*Gymnogyps califorianus*) (Fig. 5.14) preferentially feeds on large carrion. They are also known for their tendency to cache small bones in their nests, located in caves on steep cliff faces (Collins et al., 2000). Although interference with forensic investigation is possible, scavenging of human remains is relatively unlikely as the California condor is a critically endangered species with a severely limited distribution (Kiff, 2000). See Fig. 5.10.

*Figure 5.11 Black vultures (*Coragyps atratus*) (foreground). "Black vulture" by D. Mark is in the public domain.*

*Figure 5.12 Turkey vulture (*Cathartes aura*). Photograph by David Tuttle.*

Figure 5.13 Small group of turkey vultures (Cathartes aura) *feeding from an aquatic mammal carcass.* Photograph by R.T. Kramer.

Figure 5.14 California condor (Gymnogyps califorianus). *"California condor" is in the public domain.*

As obligate scavengers, vultures are highly specialized for scavenging and are quite efficient, as previously discussed in Chapter 3. Vultures are capable of significantly accelerating skeletonization, often consuming all soft tissue except for tough ligaments and skin (Ballejo et al., 2016). Evidence of vulture scavenging in soft tissue is similar to that observed

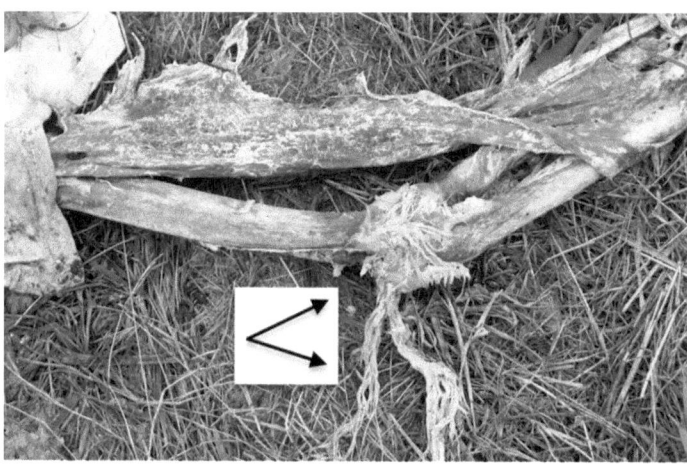

*Figure 5.15 Frayed tissue at elbow joint (*black *arrows) resulting from vulture scavenging of human remains at the Forensic Anthropology Research Facility at Texas State University. We thank Krystle N. Lewis for allowing us to photograph the remains used in her Master's thesis research.* Photograph by Susan N. Sincerbox.

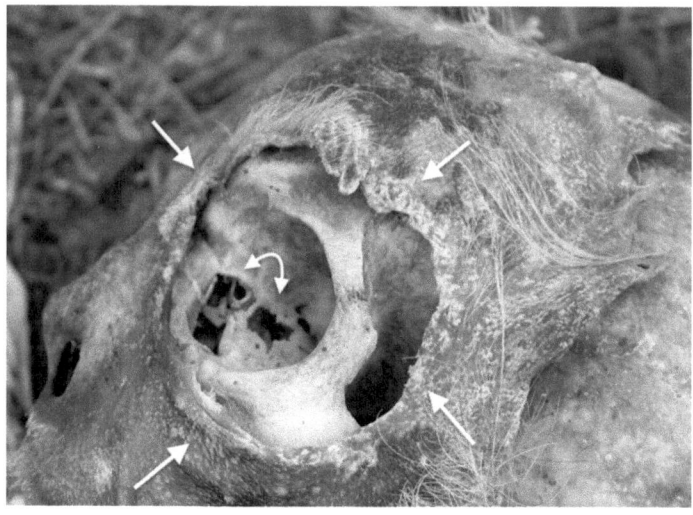

*Figure 5.16 Stretching of skin around the eye orbits (*straight *arrows) in human remains used in a vulture scaveng- ing study conducted by Krystle N. Lewis at the Forensic Anthropology Research Facility of Texas State University. Note the damage to the lacrimal and ethmoid bones inside the eye orbit (*curved *arrows).* Photograph by Susan N. Sincerbox.

in corvid scavenging, including "frayed" connective tissue and the stretching of existing orifices and wound margins (Figs. 5.15 and 5.16).

There have been several studies on the bone modifications produced by vulture scavenging, in both Old World and New World vultures

Table 5.8 Bone Modification by Vultures		
Damage Category	Traits	Reported Locations
Punctate marks	Pits[a]	Cranial bones
		Long bone diaphyses
	Punctures[a]	Cranial bones
		Scapula
Linear marks	Nonpenetrating scratches[a,b] *Changes in bone surface color, no depth*	Cranial bones Pelvic and shoulder girdles
	Shallow penetrating scoring[a,b,c] *Linear or irregular, with depth; resemble root etching*	Long bone diaphyses
Breakage	Fractures[b]	Nasals
		Lacrimals
		Ribs
		Vertebral transverse processes
	Crenulation or notching[b,c]	Scapular borders
		Mandibular condyles

[a]*Domínguez-Solera and Domínguez-Rodrigo (2011).*
[b]*Reeves (2009).*
[c]*Ballejo et al. (2016).*

(Table 5.8). Although these families evolved independently, they share similar morphology and are therefore generally comparable. Domínguez-Solera and Domínguez-Rodrigo (2011) examined modifications present on the bones of a deer fed to a population of griffon vultures (*Gyps fulvus*) in Spain. Approximately 20% of the surviving skeletal elements had at least one mark resulting from vulture scavenging. Although pits and punctures were recorded, these marks were indistinguishable from those produced by carnivore teeth, and shallow linear or irregular scoring was most common (Domínguez-Solera and Domínguez-Rodrigo, 2011). Shallow striations or microabrasion was frequently found inside the scores, mimicking modifications produced by trampling. Bone marks were concentrated on long bone diaphyses, reflecting the vulture's tendency to abandon carcasses while the cartilage surrounding the joints is still relatively intact; marks were also frequently located on the ribs, reflecting efforts to access viscera (Domínguez-Solera and Domínguez-Rodrigo, 2011).

Vultures typically begin feeding at natural orifices, often eviscerating a carcass through the anus before consuming tissues of the thoracic cavity (Beck et al., 2015; Dabbs and Martin, 2013). With vulture scavenging, the mandible is often disarticulated first, followed by the cranium and front limbs (Dabbs and Martin, 2013; Reeves, 2009; Spradley et al., 2012). Vulture scavenging can rapidly process a carcass to the skeleton, often within hours of beginning to feed, although the exact time to skeletonization will depend on the vulture species and ecological variables including wake (i.e., foraging group) size and competition with invertebrates or terrestrial scavengers. The deer placed by Domínguez-Solera and Domínguez-Rodrigo (2011) was largely skeletonized by approximately a dozen griffon vultures in under an hour of active feeding. Reeves (2009) reported variation in time to skeletonization on pig carcasses deposited in central Texas that ranged between 3 and 25 hours of active feeding, while Spradley et al. (2012) reported complete skeletonization of human remains placed at the same facility within only 5 hours of feeding time. A similar study on pig carcasses in Illinois required between 8 and 39 days to skeletonize, with most tissue loss attributed to maggots rather than vultures (Dabbs and Martin, 2013).

Vultures are also capable of picking up and transporting personal effects and skeletal elements. Such evidence may be scattered considerable distances, having a general pattern of dispersal from high to low elevation (Beck et al., 2015; Reeves, 2009; Spradley et al., 2012). In a study of vulture scavenging of pig carcasses in Arizona, Beck et al. (2015) found that after 5 weeks of exposure to vultures, the majority of elements were located within approximately 15 feet of the original sites of deposition. Ribs and vertebrae generally remained closer to their original position, while long bones were more likely to be displaced further away (Beck et al., 2015). Beck et al. (2015) suggest a minimum search area of 100 m^2, or 300 ft^2, around the original deposition site (if discernable) for vulture-scavenged remains. However, Spradley et al. (2012) found that over the course of 6 months, vultures in central Texas scattered human skeletal remains within an area of up to 900 ft^2, suggesting that a larger search area may be required if remains have been exposed for an extended period.

Cathartid Summary
- Three relevant North American species: American black vulture, turkey vulture, and California condor
- American black vultures are more likely to scavenge large carrion in groups, while turkey vultures tend toward independent or small-group foraging of smaller carcasses; California condors

unlikely to be of much forensic significance due to population vulnerability
- Turkey vultures can locate hidden carrion and can uncover shallow burials, exposing remains to other scavenging species
- Vulture scavenging of soft tissue is evidenced by stretched tissue around orifices or wounds and frayed ligaments or tendons
- Most common bone modifications are surface scratches and scoring, especially to diaphyses of meat-bearing long bones
- Vulture scavenging often accelerates skeletonization, leading to overestimations of PMI
- Vultures may scatter skeletal elements and personal effects, with elements displaced over areas as large as 900 ft^2.

FAMILY CERVIDAE—CERVIDS—DEER

Cervids, or deer (Figs. 5.17–5.22), have not been documented consuming soft tissue from carrion of any kind—not surprising given their herbivorous diets. Because they have only been documented modifying dry bone (i.e., bone in which the organic component has completely decomposed), scholars agree that they engage in bone-eating, or osteophagy, to supplement their diet with physiologically important minerals such as calcium, phosphorous, and sodium (Cáceres et al., 2011; Kierdorf, 1994). Osteophagy by cervids would be expected in populations with mineral-deficient diets, perhaps habitually in regions with mineral-deficient soils; or seasonally, in winter, when food sources are less diverse (Kierdorf, 1994; Meckel et al., 2017). Recent research has also suggested that antler chewing, a well-known form of osteophagy in cervids, may be employed to meet seasonal increases in mineral demands, such as those imposed by lactation on females at the end of the calving season (Gambín et al., 2017). The frequency of bone chewing by cervids may follow similar, demand-driven patterns. See Fig. 5.17 for the distribution of various cervid species throughout North America. The habitats, behaviors, and morphological characteristics of common North American species can be found in Tables 5.9–5.11.

Bone modification by cervids has been recognized as a phenomenon for over 30 years (Kierdorf, 1994; Sutcliffe, 1973). Similar bone-chewing behavior has also been documented in other ungulates, including wild and domestic sheep (*Ovis* spp.), camels (*Camelus* spp.), and

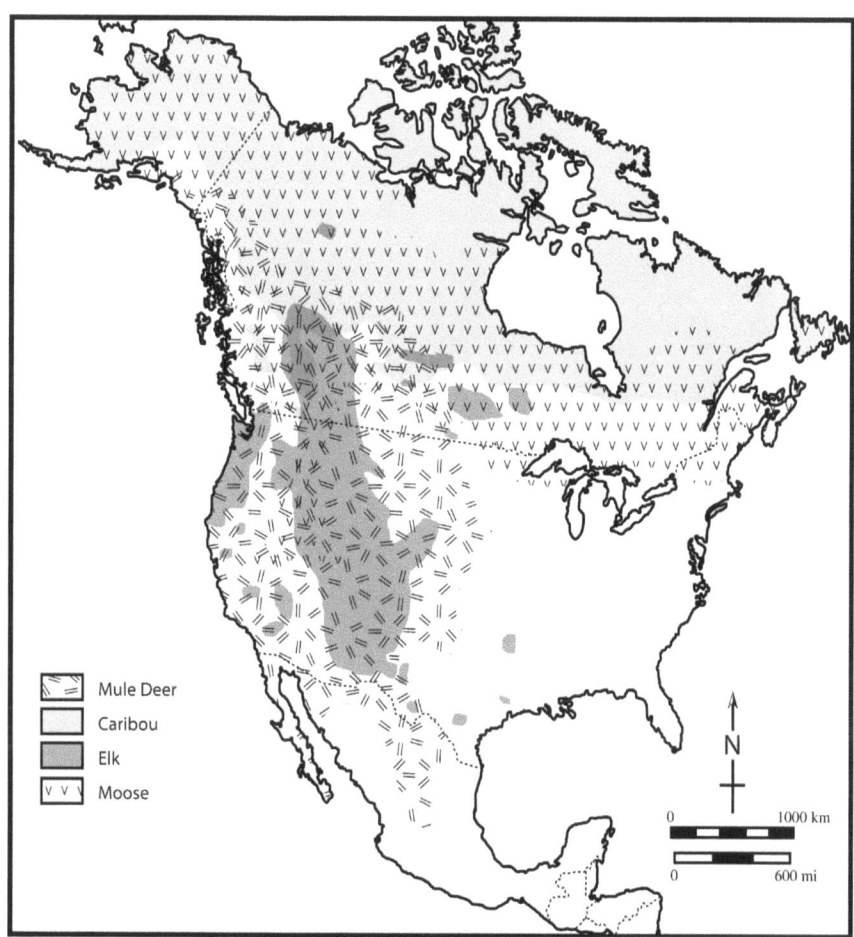

Figure 5.17 Distribution of select cervid species in North America: mule deer (O. hemionus), caribou (R. taran-dus), elk (C. canadensis), and moose (A. alces). White-tailed deer (O. virginianus), not depicted, are ubiquitous throughout the continent, except in the American southwest and far north. Distribution maps are meant only for general informational use, as species distributions change over time and individuals may occasionally be found out-side of the depicted boundaries. Distribution information compiled from Reid, F.A., 2006. The Princeton Field Guide to Mammals of North America. Houghton Mifflin Harcourt, Boston; IUCN 2017. The IUCN Red List of Threatened Species. Version 2017-1. Available from: http://www.iucnredlist.org/ (accessed 31.07.17).

giraffes (*Giraffa* spp.) (Brothwell, 1976; Hutson et al., 2013; Johnson and Haynes, 1985; Keating, 1990). However, cervid bone modification has only recently received attention by the forensic community after a game camera at the Forensic Anthropology Research Facility at Texas State University auspiciously captured a white-tailed deer (*Odocoileus virginianus*) chewing on the sternal end of a human rib (Meckel et al., 2017) (Fig. 5.23).

*Figure 5.18 Mule deer (*Odocoileus hemionus*). "Mule deer" is in the public domain.*

*Figure 5.19 Caribou, also known as reindeer (*Rangifer tarandus*). "Caribou" is in the public domain.*

Cáceres et al. (2011) identified three stages of cervid bone modification, the first of which is characterized by shallow furrows with rectangular cross sections, oriented transversely to the long bone axis and concentrated at epiphyses. Kierdorf (1994) also noted these furrows along the margins of more advanced damage. In the second stage, individual grooves are gradually obliterated as cortical bone is worn away and trabecular bone is exposed (Cáceres et al., 2011). Furrows observed in stages 1 and 2 may be mistaken for carnivore bone

*Figure 5.20 Elk (*Cervus canadensis*). "Elk" by B. Werner is in the public domain.*

*Figure 5.21 Moose (*Alces alces*). "Moose" by D. Klettke is in the public domain.*

modification, but lack the associated pits or punctures characteristic of carnivore gnawing (Cáceres et al., 2011). Furrows resulting from cervid bone-chewing also have irregular, splintered edges, are associated with microstriations in cortical bone, and may be superimposed on surfaces that have previously been weathered by the elements (Cáceres et al., 2011). These characteristics indicate that the furrows were inflicted on dry bone and are thus unlikely to be observed in cases of carnivore scavenging, where the target is the energy-rich soft tissue, grease, or bone marrow.

*Figure 5.22 White-tailed deer (*Odocoileus virginianus*). Photograph by David Tuttle.*

Table 5.9 Common North American Cervid Species		
Species Name	**Conservation Status**	**Habitat**
White-tailed deer (*O. virginianus*)	Least concern	Thrive with mixed forest cover and open areas. Abundant in eastern deciduous forests and southwestern brushland.
Mule deer (*O. hemionus*)	Least concern	Deserts, brush lands, dry forested, and mountainous regions.
Caribou[a] (*R. tarandus*)	Vulnerable	Tundra, boreal coniferous forest.
Elk (*C. canadensis*)	Least concern	Dense woodland, meadows, and plains. Migrate to higher elevations in summer.
Moose (*A. alces*)	Least concern	Tundra, northern forests, willow thickets, and swamps. Frequent bodies of water.
Conservation status courtesy of IUCN Red List of Threatened Species. [a] *Also known as Reindeer.* *Habitat information adapted from Reid, F.A., 2006. The Princeton Field Guide to Mammals of North America. Houghton Mifflin Harcourt, Boston.*		

The third stage of cervid bone modification, fork formation, is extremely diagnostic (Fig. 5.24). With one epiphysis held in the mouth, chewing by cervids wears the cortical and trabecular bone away until the marrow cavity is exposed (Cáceres et al., 2011; Kierdorf, 1994). Cervids chew side-to-side, wearing away the central portion of an epiphysis before the lateral sides. This produces a defect that is

Table 5.10 Cervid Species Introduced to North America

Species Name	Distribution
Sika deer (*Cervus nippon*)	Large populations in Texas, Maryland
Sambar deer (*Cervus unicolor*)	Small populations in Florida, Texas, and California
Axis deer (*Axis axis*)	Majority in Texas; small population on Californian coast
Fallow deer (*Dama dama*)	Small populations in British Columbia and US States of MD, GA, AL, KY, NE, TX, and CA

Distribution information adapted from Reid, F.A., 2006. The Princeton Field Guide to Mammals of North America. Houghton Mifflin Harcourt, Boston.

Table 5.11 Behavioral and Morphological Characteristics of Common Cervids

	White-Tailed Deer	Caribou	Moose
Peak daily activity[a]	Crepuscular	Diurnal	Nocturnal
Sociality[a]	Forage in small groups	Foraging groups up to 1000	Solitary; may gather while feeding
Fur[a]	Gray in winter; ruddy in summer	Dark with pale neck	Dark trunk, pale appendages
Weight (lbs) [a]	50–300	145–660	660–1300
Antlers[a]	Males only	Males and females	Males only
	Vertical points	Multiple curved branches	Palmate (flat) branches
Dental formula[b]	$\frac{0}{3} I \frac{0}{1} C \frac{3}{3} P \frac{3}{3} M$		

[a] *Reid (2006).*
[b] *Adams and Crabtree (2011).*

Figure 5.23 Photographs of a white-tailed deer chewing on a rib from skeletonized human remains at the Forensic Anthropology Research Facility at Texas State University. Photograph by the Forensic Anthropology Center at Texas State University. Reproduced with permission from Wiley. Reported in Meckel, L.A., McDaneld, C.P., & Wescott, D.J., 2017. White-tailed deer as a taphonomic agent: photographic evidence of white-tailed deer gnawing on human bone. J. Forensic Sci. DOI:10.1111/1556-4029.13514.

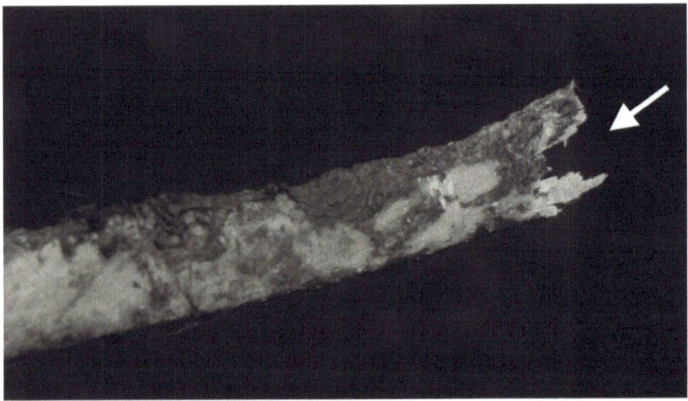

Figure 5.24 Forked defect (arrow) on the sternal end of a human rib, the result of chewing by a white-tailed deer. Photograph by the Forensic Anthropology Center at Texas State University. Reproduced with permission from Wiley. Reported in Meckel, L.A., McDaneld, C.P., & Wescott, D.J., 2017. White-tailed deer as a taphonomic agent: photographic evidence of white-tailed deer gnawing on human bone. J. Forensic Sci. DOI:10.1111/1556-4029.13514.

Table 5.12 Bone Modification by Cervids

Damage Category	Traits	Common Locations
Tooth marks	Furrowing[a,b]	Long bone epiphyses
Breakage	Gradual wear of cortical bone[b]	Long bone epiphyses
	"Forked" margins of damage[a,b,c]	Mandibular rami
		Long bone epiphyseal regions
		Sternal rib ends

[a]*Kierdorf (1994).*
[b]*Cáceres et al. (2011).*
[c]*Meckel et al. (2017).*

V-shaped or forked (Cáceres et al., 2011; Kierdorf, 1994; Meckel et al., 2017). In long bones, the midshaft tends to be well preserved with an occasional isolated furrow present (Cáceres et al., 2011). Although most common on long bones, this type of defect has also been noted on other elongated bones including the rami of mandibles and the sternal ends of ribs (Cáceres et al., 2011; Meckel et al., 2017). See Table 5.12 for characteristics of bone modification produced by cervids.

Cervid Summary

- Fairly ubiquitous distribution across the United States and Canada, and into northern Mexico

- Species in the order are herbivorous, but deer will scavenge dry bones (osteophagy) to supplement their diets with necessary minerals; populations in areas where certain minerals are naturally deficient (including winter season shortages) may be more prone to this behavior
- Tooth marks include furrows oriented transversely to the long bone axis, with conspicuous absence of pits or punctures
- Fork or V-shaped defect on epiphyseal ends, sternal ends of ribs, or mandibular rami is diagnostic

FAMILY CORVIDAE—CORVIDS—MAGPIES, CROWS, AND RAVENS

Corvids are widely distributed throughout North America, covering most of the continent (Fig. 5.25). See Tables 5.13 and 5.14 for common North American corvid species and the habitat, behavior, and morphological characteristics of each. The fact that these are common birds in addition to their omnivorous nature necessitates attention to their bone modification potential. Corvid activity is frequently reported in experimental studies of decomposition and in forensic case reports (Asamura et al., 2003; Bauer et al., 2005; Komar and Beattie, 1998; Morton and Lord, 2006; Young et al., 2014). While studies of magpie (*Pica hudsonia,* Fig. 5.26) and raven (*Corvus corax,* Fig. 5.27) scavenging strongly indicate a preference for fresh carrion, crows (*Corvus brachyrhynchos,* Fig. 5.28) have been found to scavenge throughout the stages of decomposition, but demonstrate a preference for carrion in advanced stages when increased insect activity is present (Komar and Beattie, 1998; Young et al., 2014).

Corvids target natural orifices and preexisting wounds, as even the largest species—the raven—lacks the strength to break the skin (Heinrich, 1988). Consequently, corvid scavenging tends to widen the margins of wounds and bodily orifices and distort their overall shape (Komar and Beattie, 1998). This has also been seen in cases of vulture scavenging, suggesting that the stretching and distortion of orifices or wounds may be a more general indicator of avian activity (Fig. 5.16). Corvids may also utilize wounds created by scavenging carnivores, and some species may associate with large carnivorous predators to take advantage of their abandoned kills (Heinrich, 1988; Stahler et al., 2002).

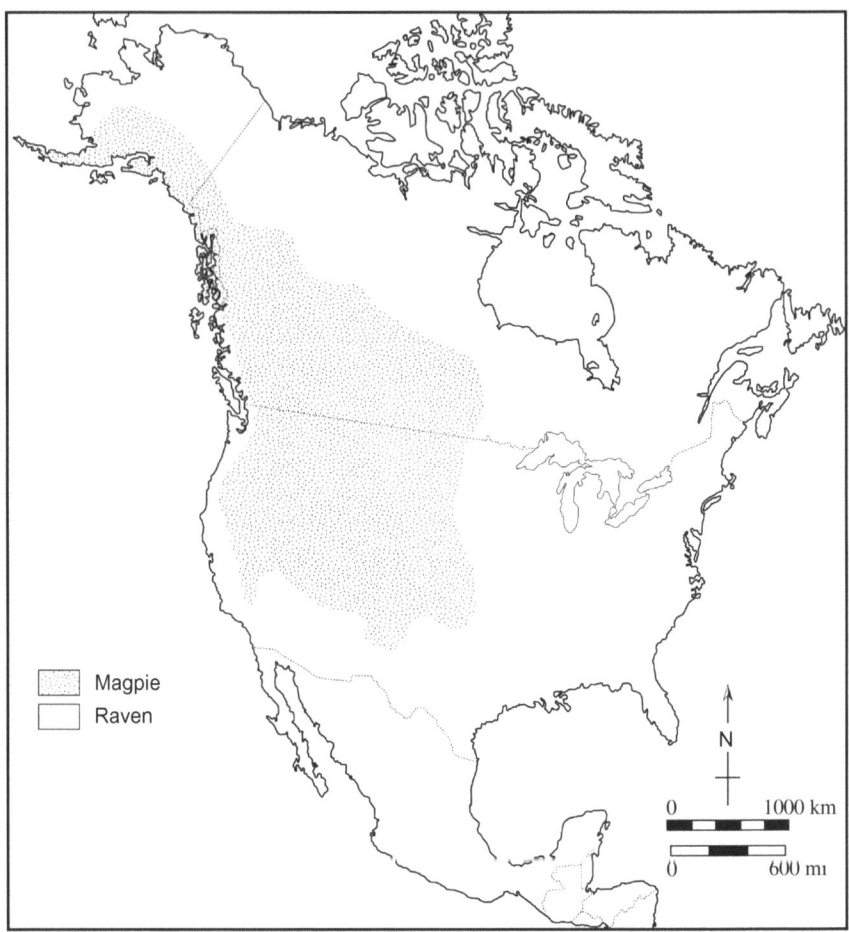

Figure 5.25 Distribution of the black-billed magpie (P. hudsonia) and common raven (C. corax) in North America. The American crow (C. brachyrhynchos), not depicted, is ubiquitous throughout the continent. Distribution maps are meant only for general informational use, as species distributions change over time and individuals may occasionally be found outside of the depicted boundaries. Distribution information compiled from Cornell Lab of Ornithology. 2015. All About Birds. Cornell University. Available from: https://www.allaboutbirds.org/ (accessed 31.07.17) and IUCN 2017. The IUCN Red List of Threatened Species. Version 2017-1. Available from: http://www.iucnredlist.org/ (accessed 31.07.17).

Table 5.13 Common North American Corvid Species

Species Name	Conservation Status	Habitat
Black-billed magpie (*P. hudsonia*)	Least concern	Meadows, grasslands, and sagebrush plains close to forested areas. Often nest near streams.
American crow (*C. brachyrhynchos*)	Least concern	Readily adapt to natural and anthropogenic environments. Preference for open woodlands. Avoids deserts.
Common raven (*C. corax*)	Least concern	Highly adaptable. Avoids open plains, uncommon in eastern forests.
Data courtesy of Cornell Lab of Ornithology. 2015. All About Birds. Cornell University. Available from: https://www.allaboutbirds.org/ (accessed 31.07.17).		

Table 5.14 Behavioral and Morphological Characteristics of Corvids			
	Black-billed Magpie	**American Crow**	**Common Raven**
Peak daily activity	Diurnal		
Sociality	Nest in pairs; forage in large groups	Roost and forage in large groups	Forage alone or in pairs; may flock at food sources
Plumage	Black, white, and blue-green	Black	Black
Weight range (lbs)	3–5	7–14	15–36
Notable characteristics	Decorative flourish on wings	Large, straight bill	Shaggy feathers at throat
	Diamond-shaped tail	Short tail	Wedge-shaped tail
Data courtesy of Cornell Lab of Ornithology. 2015. All About Birds. Cornell University. Available from: https://www.allaboutbirds.org/ (accessed 31.07.17).			

*Figure 5.26 Black-billed magpie (*Pica hudsonia*). "Magpies" is in the public domain.*

Corvids are often accompanied by other scavengers at a carcass, resulting in a general absence of data on their specific taphonomic signatures, as the overlap in carcass utilization makes it difficult to isolate species-specific modifications. Existing information on their unique effects to bone and soft tissue is summarized in Table 5.15, although it should be noted that these effects may occur due to scavenging by other avian species. For example, chickens (*Gallus gallus domesticus*) may also inflict large lesions characterized by frayed-looking tissue bridges (Roll and Rous, 1991). Further, birds of prey such as caracaras (Falconidae spp.); eagles and hawks (Accipitridae spp.); and owls (Strigiformes spp.) are also known to scavenge although their

*Figure 5.27 Common raven (*Corvus corax*). "Common raven" is in the public domain.*

*Figure 5.28 American crow (*Corvus brachyrhynchos*). "Crow" by K. Büscher is in the public domain.*

taphonomic signatures have yet to be studied (Allen and Taylor, 2013; Ballejo et al., 2016; Kane et al., 2014). While punctures in bone are general features that may be created by either avian or terrestrial species, punctures created by bills can often be distinguished from those created by teeth. Punctures created by carnivore teeth are more likely to occur in symmetrical pairs with concentrations at epiphyses, whereas punctures created by a beak are more likely to be single,

Table 5.15 Taphonomic Signatures of Corvids		
Affected Tissue	**Traits**	**Description**
Bone	Conical punctures[a]	Often occur in isolation, as opposed to in pairs
		Most often in trabecular bone
Soft tissue	Triangular punctures[a]	Shape, size consistent with beak morphology
	Large lesions[a,b]	Linear margins
		Serrated edges
	"Fraying" of tissue[a,b]	Strips of tissue with frayed appearance

[a] *Komar and Beattie (1998).*
[b] *Asamura et al. (2003).*

isolated, and found on diaphyses or on irregular bones, such as vertebrae (Komar and Beattie, 1998).

Caching of both food and inedible objects is very common in corvids, a behavioral trait that is hypothesized to have evolved from a common caching ancestor (de Kort and Clayton, 2006). Corvids are known to remove and cache small personal effects, or even bones, while scavenging. For example, Komar and Beattie (1998) observed a crow removing a credit card from an experimental carcass, with the card being recovered over 40 m from the deposition site. In the same series of experiments, the researchers recovered a rib fragment, metatarsal, and earring from a magpie nest over 600 m away. Such items are typically small, but may have significant evidentiary value. Small bones of the hands may evidence defensive sharp trauma, suggestive of a homicide; or the hyoid—the small free-floating bone in the throat—may present with fractures associated with death by strangulation or hanging. Personal effects such as identification cards, credit cards, or jewelry may expedite identification of a decedent. If scavenging by corvids is suspected, it may be worthwhile to include nearby nests in the search depending on the circumstances of the case and the resources available.

Corvid Summary

- Three common North American species: the American crow, the black-billed magpie, and the common raven
- Ubiquitous throughout North America
- Diet is omnivorous; will scavenge with a preference for fresh carrion
- Scavenging targets preexisting orifices, including any preexisting wounds

- Bone damage includes conical punctures, partially distinguishable from carnivore punctures by their singular nature
- Soft tissue damage includes lesions with straight margins and frayed edges; tissue strips left behind
- Caching is common; if feasible, nests within 600 meters should be searched for small bones and personal effects

ORDER CROCODILIA—CROCODILIANS—ALLIGATORS, CROCODILES, AND CAIMANS

Crocodilians found in North America include the American alligator (*Alligator mississippiensis*), American crocodile (*Crocodylus acutus*), and the common caiman (*Caiman crocodilus*) (Bartlett and Bartlett, 2011) (Figs. 5.29–5.32). The American alligator is found in the southeastern United States and parts of Mexico adjacent to Texas, with particularly large populations inhabiting the fresh and brackish waters of Florida and Louisiana, while the American crocodile and common caiman are found in small, isolated populations (Harding and Wolf, 2006) (Table 5.16). Small populations or isolated animals may also exist in areas where people have released their pet reptiles (Shepherd and Shoff, 2014).

Alligator attacks are fairly rare and usually result from the desensitization of the animal to humans by illegal feeding (Harding and Wolf, 2006). Although increased territoriality during the nesting season is frequently cited as a trigger for alligator aggression, studies of alligator nesting behavior suggest that female nest defense may play a smaller role than previously assumed (Harding and Wolf, 2006; Joanen, 1969; McNease and Joanen, 1981). Attacks by crocodiles are even rarer, with none reported in the United States as of 2010, as the species is wary of humans and generally less aggressive (Bartlett and Bartlett, 2011; Langley, 2010). Although smaller than the alligator or crocodile, attacks by caimans have been reported in South America where the species is more common, and caimans should not be excluded as possible predators or scavengers when completing forensic casework in regions where they are found (Haddad and Fonseca, 2011).

Given their ectothermic nature, crocodilians are more active at warmer temperatures, becoming dormant at 55 degrees Fahrenheit (Langley, 2010). Langley (2010) found that most alligator attacks in

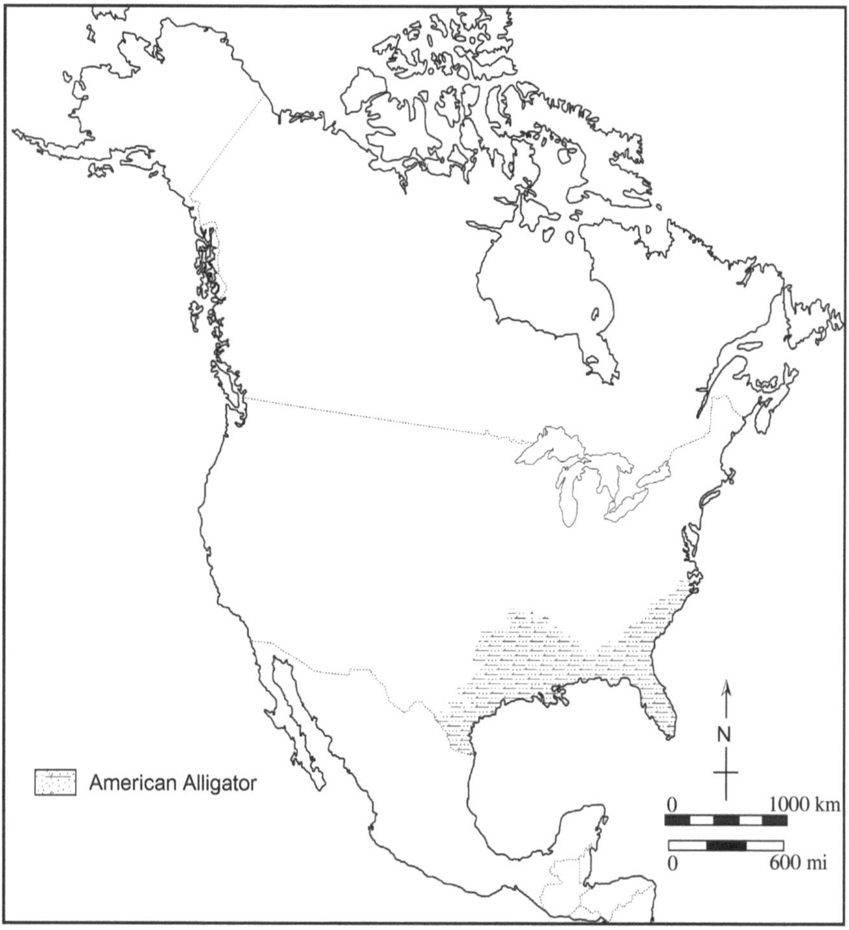

Figure 5.29 Distribution of the American alligator (A. mississipiensis) in North America. American crocodile (C. acutus) and common caiman (C. crocodilus) ranges, not depicted, overlap with American alligator distribution in southern Florida, although these species are more commonly found in Central and South America. Distribution maps are meant only for general informational use, as species distributions change over time and individuals may occasionally be found outside of the depicted boundaries. Distribution information compiled from the IUCN 2017. The IUCN Red List of Threatened Species. Version 2017-1. Available from: http://www.iucnredlist.org/ (accessed 31.07.17)

Figure 5.30 American alligator (Alligator mississipiensis). "Alligator" is in the public domain.

*Figure 5.31 American crocodile (*Crocodylus acutus*). "Crocodile" by D. Mark is in the public domain.*

*Figure 5.32 Caiman (*Caiman crocodilus*). "Spectacled caiman" by H. Braxmeier is in the public domain.*

the United States occurred between May and August, during the warmest period of the year in the northern hemisphere and thus that of peak crocodilian activity. Langley (2010) suggested that the increased likelihood of alligator attack during this period may relate to its correspondence with the alligator breeding season. Generally, the diets of wild crocodilian populations are dominated by aquatic food resources such as fish, amphibians, turtles, and aquatic invertebrates (Nifong, 2016; Platt et al., 2013; Wheatley et al., 2012). However, one study of alligator food habits found that females and juveniles incorporated higher proportions of mammalian prey in their diets, which suggests that mammalian protein sources may play important roles in reproduction and growth (Delany and Abercrombie, 1986). Further, this study cited a report by McNease and Joanen (1981) which indicated that the incorporation of mammalian protein into the diets of

Table 5.16 North American Crocodilians		
Species Name	Conservation Status	Habitat
American alligator (*A. mississippiensis*)	Least concern	Wetland ecosystems.
American crocodile (*C. acutus*)	Vulnerable	Coastal areas and river systems, preference for salt water.
Common caiman (*C. crocodilus*)*	Least concern	Lowland wetlands and riverine environments. Tolerates both salt and freshwater.

Conservation status courtesy of IUCN Red List of Threatened Species.
* Also known as the spectacled caiman.
Habitat information adapted from Bartlett, R.D., Bartlett, P.P., 2011. Florida's Turtles, Lizards, and Crocodilians. University Press of Florida, Gainesville.

female alligators improved reproductive efficiency. However, the findings of McNease and Joanen (1981) and Delany and Abercrombie (1986) have yet to be thoroughly tested. Either way, alligators may be more likely to scavenge mammalian carrion during summer months simply because of increased foraging activity, whether due to higher levels of general activity due to higher temperatures or the increased energetic demands associated with reproduction.

There is currently no published record of the taphonomic signatures produced by crocodilian scavenging of human remains in the forensic science literature. However, Bangs (2014) observed scavenging by American alligators on pig carcasses placed in the Bird Foot Delta, at the mouth of the Mississippi River in southern Louisiana. In this study, alligator scavenging was noted to begin shortly after the carcasses entered the active decay stage of decomposition, defined in this study as the exposure of internal organs below the water line. Alligator scavenging had dismembered one pig, removing it from its original location in the water, while alligators amputated the hindlimbs of a second pig carcass and extensively damaged the caudal portion of its abdomen. Complete fractures to the sternal ends of two ribs were noted in the latter pig *in situ*, but the bones could not be analyzed for other damage because unknown scavengers removed the carcass during the study period, before bones could be collected (Bangs, 2014).

Far more has been published on the characteristics of wounds inflicted by alligator attacks on living people. Harding and Wolf (2006) retrospectively document three cases of fatal alligator attacks that occurred in Southwest Florida. All three cases involved traumatic amputation, or near amputation, of distal limbs. In one case, soft tissue at the margins of an amputated lower leg was tightly coiled into tendrils (Harding and Wolf, 2006). Soft tissue tendrils may be an artifact of a behavior called the "death roll," or "twist feeding," that produces significant shear force that allows crocodilians to dismember large prey (Njau and Blumenschine, 2006; Sinton and Byard, 2016). A later report by Langley (2010) reported similar injuries in over 500 cases of adverse encounters - including both minor injuries and fatalities - with alligators in the southeastern United States. Injuries were most frequently reported on the appendages, with lacerations and conical punctures being the most frequently reported type of injury. The literature suggests that amputation or near-amputation of distal appendages is a characteristic feature of crocodilian feeding (Harding and Wolf, 2006; Langley, 2005, 2010; Sinton and Byard, 2016; Wolf and Harding, 2014). See Table 5.17 for characteristics of soft tissue injuries produced by crocodilians.

Much of the information about bone modification by crocodilians comes from the paleontological and archeological literature. From an actualistic study using captive Nile crocodiles (*Crocodylus niloticus*), Njau and Blumenschine (2006) identified two types of tooth marks that they considered to be diagnostic of crocodilian feeding: bisected tooth marks (pits or punctures) and hook scores. A bisected pit or

Table 5.17 Soft Tissue Defects Produced by Crocodilians

Damage Category	Description	Common Locations
Abrasions[a]	Linear, correspond to claw marks	Head and neck
Punctate wounds[a,b,c]	Conical, may appear in parallel rows	Head and neck
		Skin of torso, appendages
		Adjacent to amputations or tissue avulsion sites
Lacerations[a,b,c]	Elongated and irregular, with ragged edges	Skin of torso, appendages
Tissue avulsion[a,b,c]	Excision of a large portion of skin and underlying tissue, with ragged edges	Shoulder, torso, or gluteal region
		Appendages
Amputation[a,b,c]	Complete removal of a body part, at the joint or midshaft, often accompanied by tissue tendrils	Appendages
		Head or neck (decapitation)

[a] *Wolf and Harding (2014).*
[b] *Harding and Wolf (2006).*
[c] *Sinton and Byard (2016).*

puncture is divided into two by a sharp linear depression. This depression is created by raised ridges of enamel called carinae that run from the tip to the gumline of unworn crocodilian teeth (Fig. 4.6). Hook scores, the second type of diagnostic tooth mark, run a parabolic or J-shaped course over cortical bone. Hook scores have also been observed on bones modified by noncrocodilians (e.g., Komodo dragons, *Varanis komodoensis*), and have been hypothesized to be produced by species that utilize an inertial feeding strategy (Drumheller and Brochu, 2014, 2016).

Consistent with inertial feeding, crocodilian teeth and jaws are not adapted for shearing or chewing tissue but for grasping prey during a struggle. Inertial feeding involves tearing off and swallowing large chunks from a carcass, with the animal throwing its head backward and releasing the jaws. The momentum of this movement propels food backwards into the throat, where it can be swallowed. Many other reptilian species, including snakes (*Serpentes* spp.) and monitor lizards (*Varanus* spp.), have evolved such a feeding strategy (Cleuren and de Vree, 1992).

Crocodilian patterning of tooth marks also differ in comparison to that of large mammalian carnivores, with crocodilians marking fewer bones from a carcass and leaving higher density mark clusters on modified bones (Njau and Blumenschine, 2006). Drumheller and Brochu (2014) conducted an experimental study to document bone modification by the American alligator, finding that the bisected tooth marks and hook scores documented by Njau and Blumenschine (2006) are diagnostic of crocodilians in general. Additionally, all bite marks observed by Drumheller and Brochu (2014) were associated with varying degrees of crushing damage, some of which could only be observed using a scanning electron microscope (SEM). The SEM also revealed that many of the tooth marks had distinctive microstriations, likely a product of chipped carinae of worn crocodilian teeth. See Table 5.18.

In terms of tooth mark density, in areas where bone modification was apparent, Drumheller and Brochu (2014) also found a higher density of tooth marks than that which would be expected given a mammalian scavenger (with the exception, perhaps, of the kennel pattern of domestic dogs, discussed earlier). Pits and scores were most common, found on cortical bone of long bone epiphyses and diaphyses, while punctures and furrows were rare. Furrows were only found on

Table 5.18 Bone Modification by Crocodilians

Damage Category	Traits	Description	Common Locations
Bite marks	Pits[a,b]	Circular to teardrop-shaped	Epiphyses or diaphyses
		Some pits bisected	Margin of any present spiral fractures
	Scores[a,b]	Linear or slightly curved marks	Epiphyses or diaphyses
		L- or J-shaped "hook scores"	
	Punctures[a,b]	Conical	Epiphyses
		Sometimes bisected	Margin of any present spiral fractures
		Fractured margins	
	Furrows[b]	Very rare; available description is unclear	Only recorded on epiphyses
Fractures	Depressed[b,c]		Skull
			Long bones
			Ribs
	Spiral[b]		Long bone diaphyses
Other	Crushing injuries[c,d]		Skull
			Ribcage

[a] Njau and Blumenschine (2006).
[b] Drumheller and Brochu (2014).
[c] Wolf and Harding (2014).
[d] Sinton and Byard (2016).

epiphyseal regions of long bones chewed by a captive group; while punctures were more often found on epiphyses in association with depressed fractures. However, punctures were also expressed as notches near the origin of spiral fractures (Drumheller and Brochu, 2014). Bisected tooth marks and hook scores were observed more frequently in the Drumheller and Brochu (2014) study, suggesting that the alligator may feed more aggressively than the crocodile.

Crocodilian Summary

- Three relevant species: American alligator, American crocodile, and spectacled caiman
- Distribution of the alligator limited to the southeastern United States and parts of Mexico adjacent to Texas, with crocodiles and caimans found in southern Florida; isolated animals of all species may appear in areas where people have released their pet reptiles
- Carnivorous; more likely to scavenge mammalian carrion during the summer months

- Attacks by crocodiles on humans very rare; attacks by alligators and caimans relatively more common
- Soft tissue damage includes abrasions, punctate wounds, lacerations, avulsion, and amputation
- On bone, pits, scores, and punctures are common and are often bisected; L- or J-shaped hook scores as a result of inertial feeding are also frequently observed
- Bone damage includes depressed and spiral fractures, as well as crushing

FAMILY DIDELPHIDAE—*DIDELPHIS VIRGINIANA*—VIRGINIA OPOSSUM

The Virginia opossum (*Didelphis virginiana*), the only marsupial species to occupy North America, is widespread throughout the eastern and central United States, as well as along the West Coast (Reid, 2006) (Table 5.19; Figs. 5.33 and 5.34). Although their diet consists primarily of insects, the species is omnivorous and a frequent carrion consumer (Reid, 2006). In an ecological study, DeVault et al. (2011) found that opossums removed approximately 40% of mouse carcasses deposited in a heavily developed agricultural region in Indiana. Olson et al. (2016) found an even higher rate of carcass utilization by opossums, which scavenged almost two-thirds of rabbit, raccoon, and conspecific opossum carcasses deposited in a similar region. In the forensic literature, Morton and Lord (2006) reported

Table 5.19 Behavioral and Morphological Characteristics of the Virginia Opossum	
	Virginia Opossum (*D. virginiana*)
Conservation status	Least concern
Habitat[a]	Found in fields and forests, but highly tolerant of human activity. Common in agricultural, suburban, and urban areas.
Peak daily activity[a]	Nocturnal
Sociality[a]	Solitary
Fur[a]	Long hair, grizzled white and gray
Weight range (lbs)	2–15
Dental formula[b]	$\frac{5}{4} I \frac{1}{1} C \frac{3}{3} P \frac{4}{4} M$
Notable characteristics	Hairless prehensile tail
Conservation status courtesy of IUCN Red List of Threatened Species. [a] *Reid (2006).* [b] *Adams and Crabtree (2011).*	

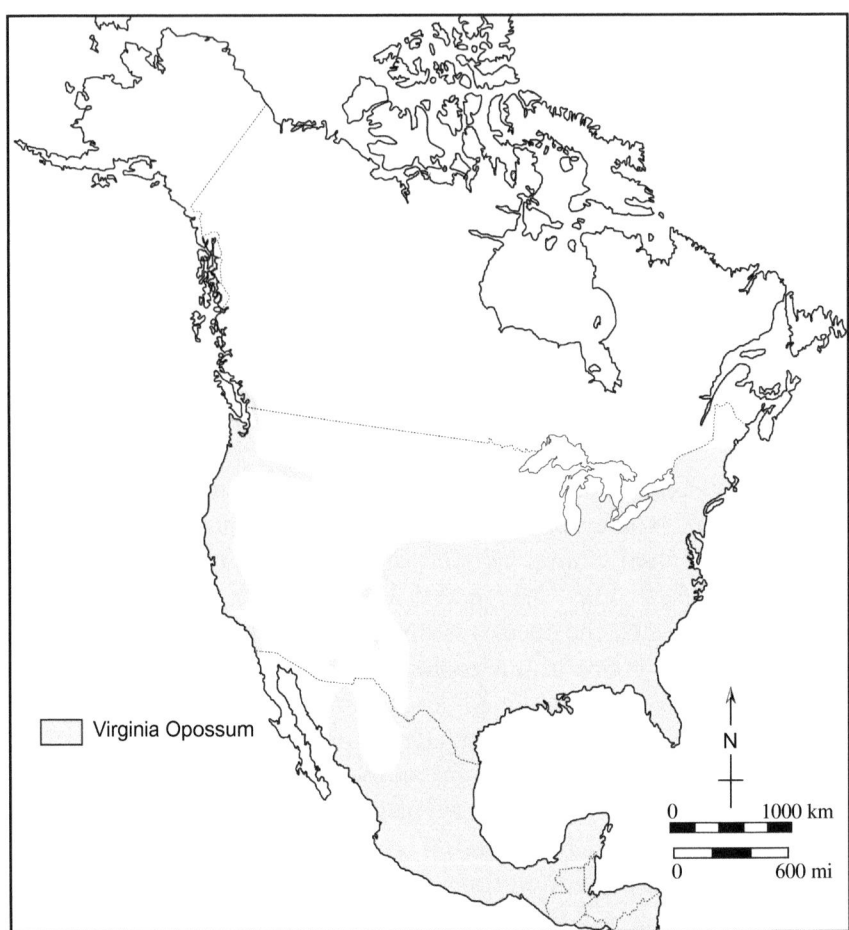

*Figure 5.33 Distribution of Virginia opossum (*D. virginiana*) in North America. Distribution maps are meant only for general informational use, as species distributions change over time and individuals may occasionally be found outside of the depicted boundaries.* Distribution information compiled from Reid, F.A., 2006. The Princeton Field Guide to Mammals of North America. Houghton Mifflin Harcourt, Boston; IUCN 2017. The IUCN Red List of Threatened Species. Version 2017-1. Available from: http://www.iucnredlist.org/ (accessed 31.07.17).

opossum scavenging of both fresh and desiccated pig carcasses in suburban Virginia. However, as opossum scavenging was observed as part of a broader study, taphonomic effects produced by opossums could not be differentiated from those produced by other scavengers such as vultures, raccoons, and foxes. Although the results of these studies cannot be completely extrapolated to forensic contexts due to the nature of the carrion items utilized, they nevertheless demonstrate that opossums are frequent scavengers in fragmented habitats (i.e., those that are divided by human infrastructure such as roads or canals).

*Figure 5.34 Virginia opossum (*Didelphis virginiana*). "Opossum on the ground" by S. Brown is in the public domain.*

King et al. (2016) produced the first report of near-exclusive scavenging by opossums. In this study, nearly 70% of all mammalian scavenger visits to pig carcasses placed in central Oklahoma were made by opossums. The authors documented only scant evidence of external tissue damage, as opossums primarily accessed internal tissues via natural orifices or postmortem defects created by other scavengers. Tooth marks on bones were similar to marks left by canids, although they were smaller and more superficial than those produced by canids (King et al., 2016). Other differences included scalloping, or repetitive semi-circular chipping, and punctures that were also present along the margins of the scapulae. Costal rib ends were splintered and split along the length of the rib, without the crushing or perpendicular breaks that are frequently documented in cases of canid scavenging (King et al., 2016). This unique pattern of breakage was attributed to the opposable thumbs of opossums, allowing them to bite and twist ribs during feeding, as observed by Morton and Lord (2006). The opossum's opposable thumb may also enable it to remove tissue and entrails by hand, similar to how raccoons scavenge, although the characteristic "hollow" or "deflated" appendages associated with raccoon scavenging have not yet been reported in opossum-scavenged remains (Jeong et al., 2016; King et al., 2016; Synstelien and Klippel, 2005). See Table 5.20 for a summary of soft tissue defects and bone modification associated with opossum scavenging.

Movement and scattering were minimal for carcasses that were scavenged only by opossums, with elements recovered within 10 m of the deposition site (King et al., 2016). However, in advanced stages of

Table 5.20 Taphonomic Signatures of Opossums		
Damage Category	Description	Reported Locations
Soft tissue damage	Skin tearing, deflation	Unspecified
Tooth marks	Small, square pits	Scapula
	Circular, superficial punctures	Scapular margins, acromion
Scalloping	Semi-circular chipping along edges	Scapular margins
Breakage	Fracturing along long axis with splintering	Costal rib ends
*Adapted from King, K.A., Lord, W.D., Ketchum, H.R., O'Brien, R.C., 2016. Postmortem scavenging by the Virginia opossum (*Didelphis virginiana*): impact on taphonomic assemblages and progression. Forensic Sci. Int., 266:576.e1-576.e6.*		

decomposition, opossums were observed to disarticulate and cache the feet and small elements such as carpals, tarsals, and vertebrae in a hollow under a nearby tree (King et al., 2016). Consequently, recovery of opossum-scavenged remains may not prove as challenging as that of remains fed on by larger scavengers. However, opossum scavenging was noted to have a confounding effect on PMI estimation. Invertebrate activity was often disrupted at scavenged carcasses, because opossums feed on both fresh tissue and insects. This serves to prolong the fresh stage of decomposition. Alternatively, opossums may also scavenge remains in advanced stages of decomposition, after peak insect activity has subsided, consuming desiccated skin and therefore exposing bone quickly (King et al., 2016; Morton and Lord, 2006). The proportion of exposed bone is a significant feature for the total body score (Megyesi et al., 2005), and therefore consumption of overlying dried tissues by opossums, which accelerates skeletonization, is likely to inflate the score and contribute to overestimation of the PMI (King et al., 2016).

Opposum Summary

- Opossums are the only extant marsupial species in North America, ubiquitous throughout the eastern and central United States, as well as part of the west coast, and most of Mexico
- Omnivorous with a preference toward insects; will scavenge carrion, both in fresh and advanced stages of decay
- Soft tissue damage not yet documented, as the species will utilize extant orifices for entry
- Bone modification includes scalloping of margins, small pits, punctures, scores, and furrows; will splinter and split rib ends
- Minimal scattering of remains, although opossums will cache small bones

- May confound PMI by consuming insects feeding on the remains, slowing progression of earlier decomposition stages; and will also feed on desiccated skin in later stages, exposing bone earlier than usual.

FAMILY FELIDAE—FELIDS—WILD AND DOMESTIC CATS

Different species of felids are widespread throughout North America (Fig. 5.35; Tables 5.21–5.22). Most are ambush predators, meaning

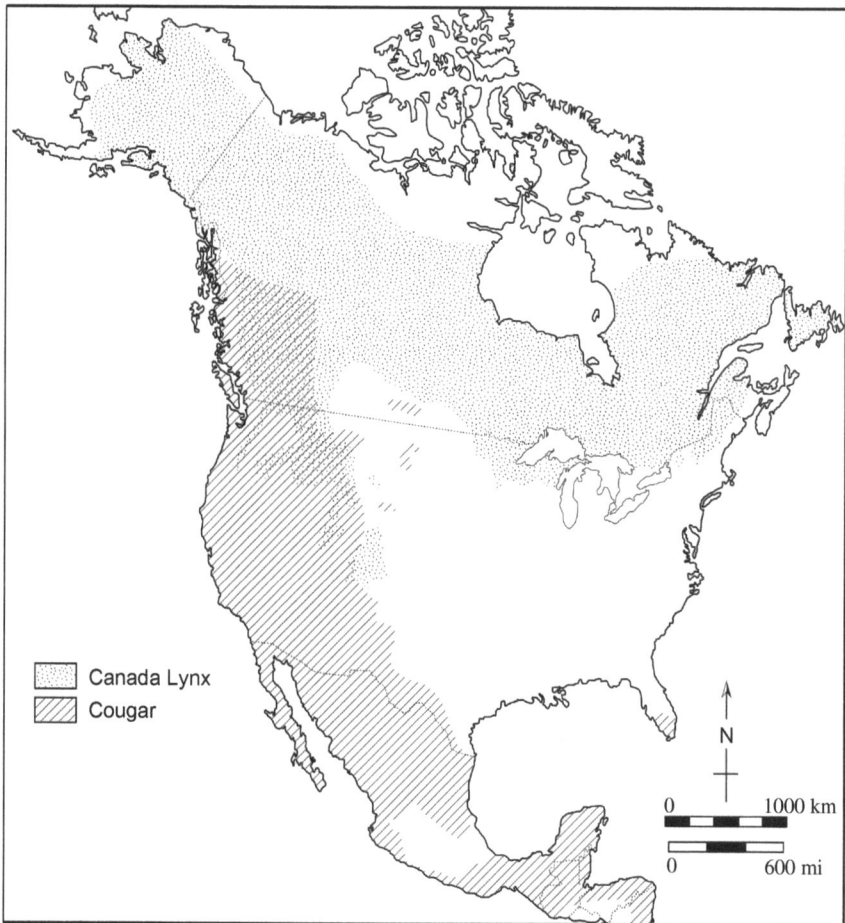

Figure 5.35 Distribution of the Canada lynx (L. canadensis*) and cougar (P.* concolor*) in North America. Bobcats (L.* rufus*), not depicted, are found throughout the continent, with the exception of some areas of the American Midwest. Domestic cats (F.* catus*), not depicted, are ubiquitous throughout the continent. Distribution maps are meant only for general informational use, as species distributions change over time and individuals may occasionally be found outside of the depicted boundaries.* Distribution information compiled from Reid, F.A., 2006. The Princeton Field Guide to Mammals of North America. Houghton Mifflin Harcourt, Boston; IUCN 2017. The IUCN Red List of Threatened Species. Version 2017-1. Available from: http://www.iucnredlist.org/ (accessed 31.07.17).

Table 5.21 Common North American Felid Species

Species Name	Conservation Status	Habitat
Bobcat (*L. rufus*)	Least concern	Mountains, deciduous or coniferous forest, swamps, arid and rocky regions.
Canada Lynx (*L. canadensis*)	Least concern	Dense coniferous forests in the north.
Puma[a] (*P. concolor*)	Least concern	Forests, mountainous regions, and deserts. Avoid water and human activity.
Domestic cat (*F. catus*)	Found throughout North America as pets, barn cats, or in feral colonies.	

Conservation status courtesy of IUCN Red List of Threatened Species.
[a]*Also known as Cougar; Mountain Lion; Florida Panther.*
Habitat information adapted from Reid, F.A., 2006. The Princeton Field Guide to Mammals of North America. Houghton Mifflin Harcourt, Boston.

Table 5.22 Behavioral and Morphological Characteristics of Common Felid Species

	Bobcat	Canada Lynx	Puma
Peak daily activity[b]	Nocturnal	Crepuscular	Nocturnal[a]
Sociality[b]	Solitary		
Coat[b]	Tawny or ruddy	Mottled gray	Sandy to red-brown
Weight range (lbs)[b]	11−40	11−37	66−265
Dental formula[c]	$\frac{3}{3} I \frac{1}{1} C \frac{3}{2} P \frac{1}{1} M$		

[a]*May be active at either time, but most often nocturnal in areas inhabited by humans.*
[b]*Reid (2006).*
[c]*Adams and Crabtree (2011).*

that they stalk prey until they are close enough to take it down in a single pounce rather than chase it down. This hunting style is heavily dependent on catching prey off guard. Consequently, energetically efficient long-distance locomotion has not been under selective pressure in felids, hindering their ability to effectively search large areas for carrion. Given the reduced probability of locating carrion, felids are rare scavengers. For example, of over 400 carcasses abandoned by puma (*Puma concolor*, more commonly referred to as "cougars" or "mountain lions" in the United States) in central Chile, most were puma kills, with only 6 being classified as cases of puma scavenging (Elbroch and Wittmer, 2013). In the forensic literature, there have been only a handful of reports of felid scavenging of human remains, including a single case of scavenging by a wild bobcat (*Lynx rufus*) (Rippley et al., 2012) and one report of scavenging by a domestic housecat (*Felis catus*) (Rossi et al., 1994) (Figs. 5.36−5.39).

*Figure 5.36 Puma (also known as mountain lion or cougar) (*Puma concolor*). "Cougar" is in the public domain.*

*Figure 5.37 Bobcat (*Lynx rufus*). "Bobcat" is in the public domain.*

When felids do scavenge, they typically consume small amounts of fresh carrion over the course of several days (Krofel et al., 2012; Rippley et al., 2012). Pumas, for example, will consume the carcasses of large prey over the course of 2–4 days (Bischoff-Mattson and Mattson, 2009). In the winter, colder temperatures reduce competition with insects and microorganisms, thus slowing decomposition and

Figure 5.38 Domestic cats (Felis catus). Photograph by Natalie R. Langley.

Figure 5.39 Canada lynx (Lynx canadensis). "Lynx canadensis" by Keith Williams is licensed under CC BY 2.0.

preserving fresh tissues for extended periods of time. Such higher levels of preservation may also extend the scavenging window for felids.

Experimental research that simulated puma caches has demonstrated that felid caching improves carcass preservation, thus extending the scavenging window (Bischoff-Mattson and Mattson, 2009). Felids will cache carcasses by covering them with nearby soil, vegetation, or snow (Vander Wall, 1990). Although felids do not dig to create caches, as canids sometimes do, they may move carcasses to natural depressions, dense vegetation, or shade before covering them (Bauer et al., 2005; Bischoff-Mattson and Mattson, 2009; Vander Wall, 1990). Linear striations may be found in the soil surrounding the cache, where the paws have been used to scrape debris over the carcass (Rippley et al., 2012). Often, the carcass's own hair or fur will be incorporated in the cache material, a behavior not observed in caches constructed by other large carnivores (Rippley et al., 2012; Vander Wall, 1990). Out of the felid species in North America, pumas are most likely to cache remains (Pokines, 2014).

Given the rarity of its occurrence, there has been little research on the taphonomic effects of felids on soft tissue and bone. Large felids have been proposed to target viscera first, opening the thoracic cavity (Pickering, 2001). Consequently, scavenging damage may be concentrated at the trunk. In a series of studies, researchers fed baboon (*Papio cynocephalus*) carcasses to captive leopards (*Panthera pardus*) (Pickering, 2001; Pickering and Carlson, 2004). Damage was concentrated at skeletal elements of the thoracic cavity and abdomen such as the ribs, clavicles, scapulae, vertebrae, and innominates (Pickering, 2001). They may also target the hands and feet, often swallowing digits, carpals, and tarsals whole (Pickering, 2001; Pickering and Carlson, 2004). Consequently, entire digits can sometimes be recovered from scat with minimal damage. However, the application of this research to forensic casework warrants further investigation due to behavioral differences that may arise given the use of a primate analog and captive felid population.

The pattern of tissue consumption may vary under certain circumstances. In one case, a jaguar (*Panthera onca*) at an animal sanctuary in North Carolina was observed targeting the head of a deer carcass placed in its enclosure, even though the deer had extensive open wounds in the abdomen (N. Siegel, personal communication, August 3, 2017). Absence of the viscera due to an autopsy affected felid consumption in the only published report of bobcat scavenging of human remains in North America (Rippley et al., 2012). Rather than concentrating on the thoracic cavity, the bobcat targeted muscle tissue in the arms, hips, and thighs. The small size and reduced strength of domestic cats, which prevents opening of the thoracic cavity, may also result in deviations from the consumption patterns reported in large felids. In a case reported by Rossi et al. (1994), a man was found dead at home with flesh missing from his head, throat, neck, and right upper arm.

Cause of death was determined to be related to a drug overdose and the postmortem damage was attributed to the decedent's domestic cats (Rossi et al., 1994).

The observations of a bobcat feeding on human remains discussed earlier were obtained during a decomposition study conducted in southeast Texas (Rippley et al., 2012). Scratches were present on the abdominal region and wound margins in the soft tissue were smooth, resembling incised wounds (i.e., those produced with a sharp instrument). The only noted skeletal damage was to the left forearm, with punctures on the distal epiphysis of the ulna and destruction of the head of the second metacarpal. There was no evidence of epiphyseal gnawing for grease extraction, which is common in cases of canid scavenging. As decomposition advanced, scavenging by vultures destroyed the scant evidence of the bobcat's feeding (Rippley et al., 2012). Similar destruction of evidence by other scavenging species, together with the general rarity of felid scavenging, may help to explain the absence of reported felid scavenging in forensic cases. See Table 5.23 for a summary of felid taphonomic signatures in soft tissue and bone.

Table 5.23 Taphonomic Signatures of Felids		
Affected Tissues	**Traits**	**Reported Locations**
Soft tissue	Scratches[a]	Abdomen
	Incised wound margins[a]	Margins of any soft tissue damage
Bone	Tooth marking[a,b,c,d,e] (Pits, punctures, scoring)	Scapula
		Long bones
		Innominates
	Multi-cusp punctures[d,e]	Unspecified
	Scalloping[c,d,e]	Mandible
		Scapula
		Innominates (especially ilium)
	Fractures (irregular, transverse, oblique)[b,d]	Ribs
		Long bones
	Destruction of features[a,c,d,e]	Mandibular condyle
		Vertebral processes
		Olecranon process of ulna
		Head of metacarpal
[a]*Rippley et al. (2012).*		
[b]*Moran and O'Connor (1992).*		
[c]*Álvarez et al. (2012).*		
[d]*Kaufmann et al. (2016).*		
[e]*Mondini and Muñoz (2008).*		

Scavenging (or Lack Thereof) by the Domestic Cat

A single report of scavenging by domestic housecats has been published in North America, inferred from postmortem soft tissue damage and the presence of the decedent's cats (Rossi et al., 1994). Neophobia, or the fear of new foods, may contribute to the avoidance of certain housecats to raw meat and, thus, to scavenging carrion (Bradshaw et al., 2000). Feral domestic cats or barn cats, however, may be more prone to scavenging as a greater portion of their normal diet consists of raw meat (Bradshaw et al., 2000). There has been one experimental study of bone modification by domestic cats, which observed that bone punctures were particularly deep, narrow, and often paired, consistent with feline molar morphology (Moran and O'Connor, 1992). Like other carnivores, damage was heavily concentrated at the epiphyses of long bones and the margins of scapulae. Unfortunately, the bones involved in the study were introduced in an unrestricted outdoor context and scavenging was not supervised, and thus the bone modifications cannot be conclusively attributed to domestic cats alone (Moran and O'Connor, 1992).

Although research on scavenging by the domestic cat is scarce, bone modification patterns associated with heavy scavenging may be cautiously inferred from a similarly-sized South American felid species, Geoffroy's cat (*Leopardus geoffroyi*). Álvarez et al. (2012) examined patterns of bone modification of rabbit carcasses by a captive population of these cats. While small skeletal elements (e.g., bones of the hands and feet, vertebrae) were recovered intact with few tooth marks, most other bones were fragmentary and presented with irregular, transverse, or oblique fractures (Álvarez et al., 2012). Overall, tooth marking was relatively rare, noted in less than 20% of the recovered elements (Álvarez et al., 2012). Marked bones presented with pits, punctures, and scoring, most commonly found on the innominates, scapulae, and long bones; scalloping was also common in flat and irregularly shaped elements, such as the innominate, scapula, and mandible (Álvarez et al., 2012). Overall, scavenging of *small* mammal carcasses by Geoffroy's cat leaves patterns consistent with the taphonomic signatures observed for larger felid species. In larger carrion, bone modification by domestic cats would likely be limited to the smaller bones of the hands and feet and may resemble damage patterns seen in the long bones of rabbits.

Felid Summary
- Canada lynx are ubiquitous across Canada and Alaska, with smaller populations in the northern and western United States; Pumas reside in western Canada and the United States and throughout Mexico, with a population in Florida; Bobcats are found through most of the continent with the exceptions of the far north, American midwest, and southern Mexico; Domestic cats are ubiquitous across the continent
- Felids are carnivores yet rare scavengers, unlike canids
- Scavenging by large felids likely to target the thoracic cavity
- Caching will occur by concealing remains at the surface, rather than burying them
- Bone damage includes pits, punctures, and scoring, although only a minority of bones exhibit damage

FAMILY PROCYONIDAE—
PROCYON LOTOR—NORTHERN RACCOON

The Northern Raccoon (*Procyon lotor*) is an omnivorous species and frequent scavenger of carrion (Fig. 5.40). In part due to its dietary flexibility, raccoons are abundant throughout North America and absent only from the northernmost regions of Canada (Reid, 2006). Raccoons prefer habitats characterized by hardwood forest, especially those adjacent to bodies of water (Reid, 2006; Synstelien and Klippel, 2005).

*Figure 5.40 Raccoon (*Procyon lotor*). "Raccoon" by O. Lenz is in the public domain.*

Table 5.24 Behavioral and Morphological Characteristics of the Northern Raccoon	
	Northern Raccoon (*P. lotor*)
Conservation status	Least concern
Habitat	Variable, but preference for wetlands and forests. Highly tolerant of human activity and often found in agricultural, urban, or suburban areas.
Peak daily activity[a]	Nocturnal
Sociality[a]	Solitary
Fur[a]	Long, mottled gray
Weight range (lbs)[a]	5–33
Dental formula[b]	$\frac{3}{3}I\ \frac{1}{1}C\ \frac{4}{4}P\ \frac{2}{2}M$
Notable characteristics	Black mask and banded black-and-white tail[a]
	Pentadactyl (have five flexible digits on each limb)[c]

[a]*Reid (2006).*
[b]*Adams and Crabtree (2011).*
[c]*Hannigan (2015).*

However, raccoons are extremely adaptable to alternative habitats and are well-known fixtures in urban environments, subsisting on human refuse (Reid, 2006; Synstelien and Klippel, 2005) (Table 5.24).

Although raccoons are frequent scavengers, their contributions to taphonomy have only recently been addressed by the forensic community. Most research on raccoon scavenging has occurred at the Anthropology Research Facility (ARF) of the University of Tennessee, Knoxville (Jeong et al., 2016; Smith, 2015; Synstelien, 2013; Synstelien and Klippel, 2005). Raccoons are the most commonly reported scavenger at the facility, affecting over half of the donated human remains placed between 2011 and 2013 (Jeong et al., 2016; Steadman et al., 2016). The high prevalence of raccoon scavenging at ARF, as opposed to other human decomposition research facilities, is likely a consequence of raccoon habitat preferences and the facility's proximity to the Tennessee River (Smith, 2015; Synstelien and Klippel, 2005).

Raccoons demonstrate a preference for fresh tissue, with all raccoon scavenging recorded at ARF occurring prior to the onset of the bloat stage of decomposition (Jeong et al., 2016; Smith, 2015). Although raccoons avoid consuming putrefying tissue, they have been demonstrated to forage for the invertebrate larvae which colonize remains during more advanced stages of decomposition (Jeong et al., 2016; Smith, 2015; Synstelien and Klippel, 2005). This disruption of invertebrate colonization, combined with increased air flow over soft tissue through points of entry created by scavenging raccoons, appears to improve the conditions for tissue desiccation and increases the likelihood of partial or complete mummification, which could impact PMI estimates (Jeong et al., 2016; Smith, 2015). In a study at ARF, Jeong et al. (2016) reported that approximately 75% of raccoon-scavenged remains mummified, compared to a mere 35% of unscavenged remains placed during the same time period. Although all raccoon scavenging at ARF has been reported in surface deposits, raccoons have also been observed to locate and expose shallow burials of small pig carcasses in Virginia (Morton and Lord, 2006). Consequently, investigators should not exclude raccoon scavenging from consideration when interpreting disturbed burials, especially shallow ones.

In Tennessee, raccoon scavenging on human remains placed at ARF was more prevalent during the summer, peaking during July (Jeong et al., 2016). A study from Maine illustrates a different pattern, reporting that the highest prevalence of raccoon scavenging on pig carcasses occurred during late winter and early spring, peaking in March (Hannigan, 2015). This period corresponds with raccoon mating season in Maine, and the author of the study suggests that the discrepancy may be due to differences in the timing of the mating season between northern and southern states (Hannigan, 2015). However, the mating season for raccoons in Tennessee also begins in February, and most pups are born long before the peak scavenging prevalence in July. A recent study at ARF offers an alternative explanation of the existing difference in peak scavenging season between raccoons in Tennessee and Maine. In a study of vertebrate scavenging using different carcass types, Steadman et al. (2016) found that raccoons scavenged human remains throughout the year but only showed interest in pig carcasses during the winter months. Consequently, the conflicting results of the study in Maine may be a mere artifact of using pigs as a human proxy, demonstrating the importance of conducting forensic decomposition research using actual human remains rather than animal analogs when possible.

In addition to eating soft tissue, raccoons may scavenge tougher cartilaginous facial tissues, including the ears (Smith, 2015). However, one taphonomic signature produced by raccoon scavenging is highly diagnostic. Unlike many other vertebrate scavengers, raccoons do not appear to be interested in visceral or abdominal tissue, even if it is easily accessible. Instead, they target the muscular tissues of the appendages. Their method of feeding on this tissue is equally unique. Raccoons create points of entry by chewing small holes in the skin, often at joints, such as the elbow or knee. The hands and feet are also frequent entry sites. Raccoons then use their forepaws to extract muscle tissue through this hole for consumption, which often stretches the skin and widens the original entry point. Skin in these areas may exhibit localized slippage or superficial scratches from the claws (Smith, 2015). This behavior empties the limb of muscular tissue, including the gluteal and shoulder regions, resulting in a skin-covered limb that has a collapsed or deflated appearance (Jeong et al., 2016; Smith, 2015).

However, the first author has observed that a "deflated" limb can also be created without activity by a vertebrate scavenger, where heavy maggot activity is restricted to internal bodily cavities due to intense solar radiation. Furthermore, scavenging by vultures and corvids can also create stretched openings in the skin, although these will typically be found on the face or abdomen; when openings created by bird scavenging are found on the limbs, they will often be accompanied by tissue fraying (Komar and Beattie, 1998). Attributing postmortem damage to raccoon scavenging should be done with caution, giving attention to evidence of other possible scavenger activity as well as the environmental context in which the remains were found.

Scant research has been conducted on the taphonomic effects of raccoon scavenging in bone. Hannigan (2015) notes that raccoon scavenging was concentrated on soft tissues of pig carcasses, with little evidence of bone modification. A caveat to the findings from the Hannigan (2015) study, as noted above, is that it was conducted using nonhuman animals as proxies. A study of raccoon scavenging on human remains by Synstelien (2013) observed that minor bone modification could occur in the course of soft tissue consumption. As with other carnivorous scavengers, tooth mark morphology includes pits, punctures, scores, and furrows. However, raccoon tooth marks are less pronounced and include more crushing damage than those produced by larger carnivores due to the reduced nature of raccoon carnassial teeth (Synstelien, 2013). Raccoon scavenging in skeletal remains is also distinguished from scavenging by other carnivores by reduced destruction to long bone epiphyses and substantially greater damage to bones of the hands, feet, ribs, and vertebrae (Synstelien, 2013). Both Hannigan (2015) and Synstelien (2013) report minimal scatter of

Table 5.25 Taphonomic Signatures of Raccoons		
Affected Tissues	**Traits**	**Reported Locations**
Soft tissue	Consumption of cartilage[a]	Facial tissues (e.g., ears, nose)
	Entry point with stretching, localized skin slippage, and superficial scratches[a,b]	Hands/feet
		Joints (e.g., elbow, knee)
	Removal of muscle tissue ("hollowed limbs")[a,b]	Arms/shoulder region
		Legs/gluteal region
Bone	Superficial pits, punctures, scores, and furrows[c]	Concentrated at hands, feet, ribs, and vertebrae

[a]Smith (2015).
[b]Jeong et al. (2016).
[c]Synstelien (2013).

skeletal elements by raccoons. See Table 5.25 for a summary of the taphonomic signatures produced by raccoons.

Raccoon Summary

- Raccoons are ubiquitous throughout North America and prefer woodland habitats near water, but are also well adapted to urban environments
- Omnivorous and frequent scavengers of carrion
- Preference for fresh carrion, avoids carrion in bloat stage and later
- Will consume invertebrate larvae, which may enhance conditions leading to mummification and skewed PMI estimates
- Prefer to scavenge surface deposits, but scavenging shallow burials is not unheard of
- Soft tissue consumption is unique: cartilaginous areas of the face may be targeted, and entry into the body is at the limb joints, where the animals remove muscle mass for consumption, resulting in a deflated limb appearance. Skin at entry points may exhibit superficial scratches or localized slippage
- Such soft tissue signs should be only attributed to raccoons depending on context, as other scavengers (i.e., vultures, corvids) may leave behind similar marks
- Bone damage includes pits, scores, punctures, and furrows but tooth marks are less pronounced and are not as common as those associated with larger carnivores
- Bone damage to hands, feet, ribs, and vertebrae more extensive than that associated with other carnivores

ORDER RODENTIA—RODENTS

The diversity of the order Rodentia is substantial, with almost 2000 species worldwide (Reid, 2006). Abundant throughout North America, taxa include New World rats and mice (Muridae spp.), chipmunks and tree squirrels (Sciuridae spp.), and introduced European rats (also Muridae spp.); and are regularly reported to have taphonomic effects on human remains (Figs. 5.41–5.45). Additional rodent taxa found in

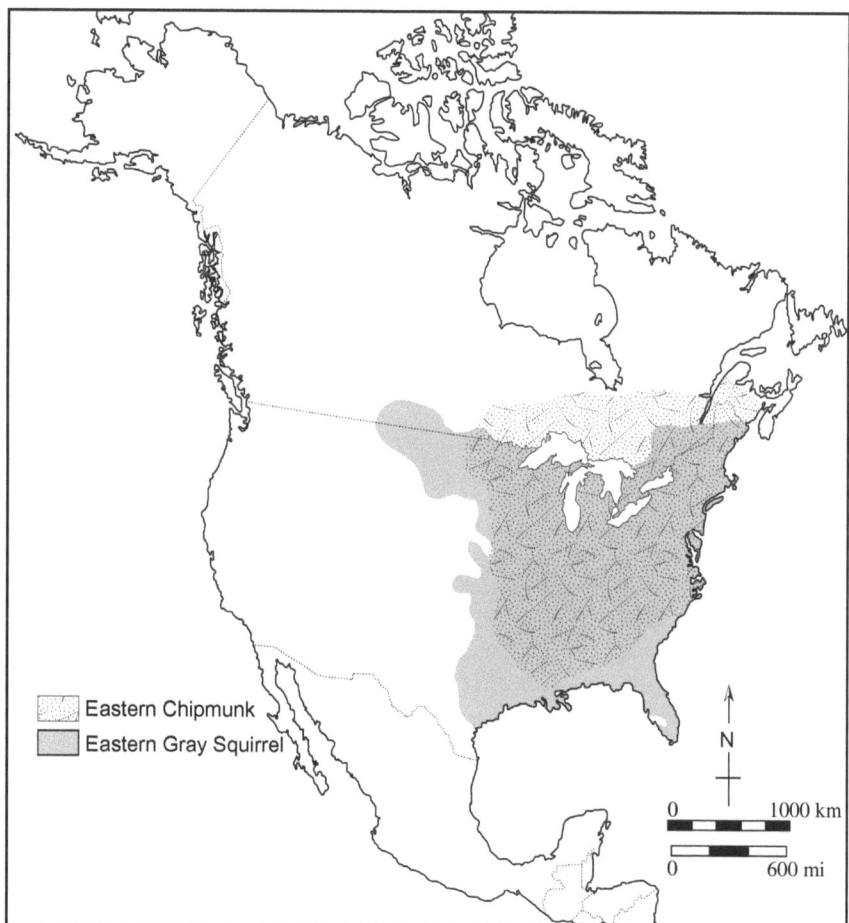

Figure 5.41 Distribution of chipmunks (T. striatus) and eastern gray squirrels (S. carolinensis) in North America. Distribution maps are meant only for general informational use, as species distributions change over time and individuals may occasionally be found outside of the depicted boundaries. Distribution information compiled from Reid, F.A., 2006. The Princeton Field Guide to Mammals of North America. Houghton Mifflin Harcourt, Boston; IUCN 2017. The IUCN Red List of Threatened Species. Version 2017-1. Available from: http://www.iucnredlist.org/ (accessed 31.07.17).

*Figure 5.42 Eastern chipmunk (*Tamias striatus*). "Chipmunks" by E. Ochel is in the public domain.*

*Figure 5.43 Eastern gray squirrels (*Sciurus carolinensis*). Photograph by Elizabeth A. DiGangi.*

North America include ground squirrels and prairie dogs (Sciuridae spp.); lemmings and voles (Arvicolinae spp.); rabbits and hares (Leporidae spp.); beavers (*Castor* spp.); and porcupines (Erethizontidae spp.) (Reid, 2006). Many rodent species are highly tolerant of human activity and are often found domestically, as either pets or pests. These

*Figure 5.44 Brown rat (*Rattus norvegicus*). "Rat surmulot/Brown rat" by Jean-Jacques Boujot is licensed under CC BY-SA 2.0.*

*Figure 5.45 Mouse (*Mus musculus*). "Mouse" is in the public domain.*

domestic species have also been reported to scavenge human remains in indoor contexts (Ropohl et al., 1995; Tsokos et al., 1999; Tsokos and Schulz, 1999). When identifying a scavenging rodent, an inclusive resource on rodent taxa should be consulted, but below we provide information regarding habitat preferences, behavior, and morphological characteristics for four common species (Table 5.26).

Table 5.26 Behavioral and Morphological Characteristics of Representative Rodent Species				
	House Mouse (*M. musculus*)	Eastern Gray Squirrel (*S. carolinensis*)	Eastern Chipmunk (*T. striatus*)	Brown Rat (*R. norvegicus*)
Conservation status	Least concern	Least concern	Least concern	Least concern
Habitat[a]	Widespread, rural and urban areas	Widespread, rural and urban areas	Deciduous forest, suburban areas	Widespread, urban areas
Peak daily activity[a]	Nocturnal	Diurnal	Diurnal	Nocturnal
Nesting behavior[a]	Ground burrows	Arboreal nests	Ground burrows	Ground burrows
Caches?[a]	Unknown	Yes, scatter	Yes, larder	Unknown
Weight (oz)[a]	Approx. 0.5	12–24	Approx. 3.5	Approx. 14
Mean incisor width (upper/lower, in mm)[b]	0.42/0.35	1.56/1.40	0.98/0.96	1.23/1.12

Conservation status courtesy of IUCN Red List of Threatened Species.
[a]Reid (2006).
[b]Pokines et al. (2017).

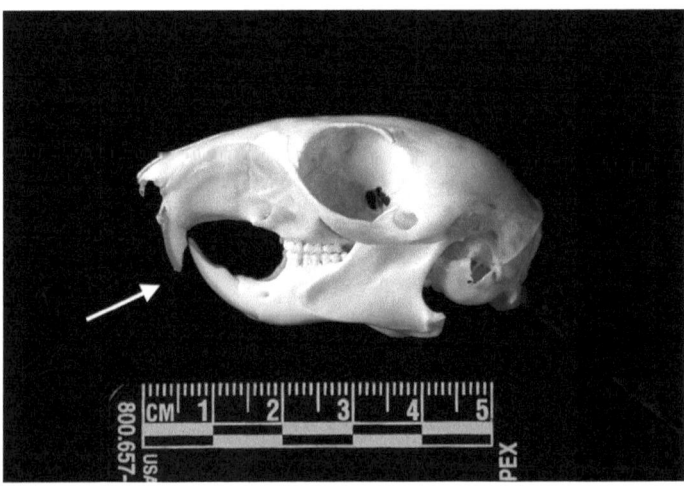

Figure 5.46 Skull of Eastern gray squirrel (S. carolinensis). *Note the prominent incisors.* Photograph by David Tuttle.

Rodents vary tremendously in size, with some extant species weighing in as little as 7.0 g (e.g., pygmy mouse, *Baiomys taylori*) or as much as 60.0 kg, over 130 lbs. (capybara, *Hydrochoerus hydrochaeris*) (Kay and Hoekstra, 2008; Reid, 2006). The most distinctive morphological characteristic of rodents is their incisors. Rodents have a single pair of ever-growing incisors with chisel-shaped occlusal surfaces on their upper and lower jaws (Kay and Hoekstra, 2008; Reid, 2006) (Fig. 5.46). The width

Table 5.27 Taphonomic Signatures of Rodents		
Affected Tissues	**Traits**	**Description**
Soft tissue	Wound margins	Scalloped or serrated edges[a,b,c,d,e]
		Parallel cutaneous lacerations along wound margin[b,d,f]
		Skin surrounding wound undamaged[a,c,d,e]
	Defects	Rounded defects[d]
		Layered destruction of skin; crater-like tissue loss[a,b,c]
Bone	Punctures	Square or rectangular[f]
	Furrows	Flat-bottomed, shallow[a,f,g]
		Parallel series of paired furrows from gnawing incisors[a,c,f,g]
	Destruction of features[*]	Layered removal of trabecular bone, resulting in "shaved" appearance[f]
		Destruction of trabecular bone leading to hollowed marrow cavities[f]
[a]*Haglund (1992).*		
[b]*Ropohl et al. (1995).*		
[c]*Haglund (1997b).*		
[d]*Tsokos and Schulz (1999).*		
[e]*Tsokos et al. (1999).*		
[f]*Klippel and Synstelien (2007).*		
[g]*Pokines (2015).*		
[*]*Seen predominantly in scavenging by rats.*		

of the incisor varies from species to species and is largely responsible for the morphology of furrows produced by rodent gnawing (Haglund, 1997b). Although the dimensions of furrows may inform the general size class of a rodent scavenger, rodent species are most often identified based on faunal evidence, particularly excrement or hair (Pokines, 2015; Tsokos et al., 1999).

Rodents have varying motivations for scavenging. Most species gnaw on bones to maintain the length of their ever-growing incisors and obtain minerals, while others (predominantly rats) are interested in soft tissue for its value as a nutritious energy source (Klippel and Synstelien, 2007). In the remainder of this section, these species will be referred to as either mineral-motivated or calorie-motivated rodents. The taphonomic signatures (Table 5.27) of mineral-motivated and calorie-motivated rodents are described in the following paragraph, which summarizes a study conducted by Klippel and Synstelien (2007) in eastern Tennessee.

Mineral-motivated rodents will only gnaw bone when it is "dry," or devoid of fat and grease. The amount of time required for bone to dry depends largely on the decomposition environment, occurring faster when bones are deposited in full sun than if deposited in shade. Mineral-motivated species will target skeletal regions with thick cortical bone, which can be shaved away and consumed. While calorie-motivated rodents may still gnaw bone to maintain incisor length, producing characteristic furrows to cortical regions, they also modify bone while consuming soft tissue. Consumption of soft tissue by calorie-motivated rodents focuses on exposed regions of the body, often concentrating on the soft tissues and cartilage of the sensory organs (Tsokos et al., 1999). Brown rats produce such damage, demonstrating a preference for tissues with high fat content. In bone, brown rats typically target articular cartilage and skeletal regions with thin cortical bone that can be easily broken through to access the grease-laden trabecular bone underneath.

In soft tissue, rodents produce rounded defects with scalloped or serrated edges (Tsokos and Schulz, 1999). Tissue is often excised in layers, starting at a central point and working outward, resulting in crater-like defects (Haglund, 1992, 1997b; Ropohl et al., 1995). The skin surrounding the defect is often undamaged, but parallel lacerations may occasionally be present at the margins (Haglund, 1997b; Tsokos et al., 1999). In bone, punctures may occur. Furrows, however, are more common (Fig. 5.47). These furrows have flat bottoms, are

Figure 5.47 Butchered large nonhuman mammal bone modified by (likely large) rodents. Note characteristic parallel furrows along the margins of the weathered shaft that often superimpose one another. The bone fragment is approximately 3 × 5 cm. Photograph by David Tuttle.

relatively shallow, and are often found in pairs (Haglund, 1997b; Klippel and Synstelien, 2007; Pokines, 2015). In scavenging by calorie-motivated rodents, trabecular bone may be destroyed in layers, giving it a "shaved" appearance (Klippel and Synstelien, 2007). Ultimately, epiphyses may be worn away entirely in cases of scavenging by calorie-motivated rodents, leaving diaphyses devoid of marrow (Klippel and Synstelien, 2007).

Rodents also engage in a number of unique behaviors which result in modification or displacement of remains. Rodents have been reported to use mummified or skeletonized human remains as nesting locations, sometimes introducing nesting materials such as plant matter or fibers to the body (Haglund, 1992, 1997b; Pokines, 2015). Although rodents are anecdotally known to nest in the thoracic cavity, one report documented recovery of the nest of a woodland vole (*Microtus pinetorum*) from inside a human skull (Pokines, 2015). However, in many cases, occupation of the remains may not be possible due to colonization by invertebrates or disruption by larger scavengers.

Many rodent species are ardent cachers and may transport tissue or small skeletal elements in specialized cheek pouches, bringing the finds to nearby burrows or nests (Haglund, 1992; Ropohl et al., 1995). In one unique case, a free-ranging pet hamster (*Mesocricetus auratus*) moved small particles of human tissue to a nest it built in the home following the death of its owner (Ropohl et al., 1995). Other rodent species are scatter-hoarders, and may create many small, dispersed caches in preparation for seasonal food shortages (Vander Wall, 1990).

Rodent Summary

- Hundreds of rodent species found throughout North America; responsible scavenging species often identified using faunal evidence
- Rodents may use carrion as a source of energy (calorie-motivated) or essential minerals (mineral-motivated); skeletal damage by mineral-motivated rodents will be concentrated on thick cortical bone, while damage by calorie-motivated species will focus heavily on regions rich with trabecular bone
- Soft tissue defects produced by rodents are rounded, layered, have scalloped edges, and are surrounded by unbroken skin
- Series of parallel, paired furrows are the most common modification observed in rodent-scavenged bones

- Rodents may occupy the remains, introducing nesting materials to the thoracic cavity or skull
- Many species of rodents are scatter-hoarders and may cache tissue or small skeletal elements

SUPERORDER SELACHIMORPHA—SHARKS

Various species of shark are found in populations all along the coastline of North America (Figs. 5.48 and 5.49). From a forensic standpoint, identifying sharks and interpreting their taphonomic signatures is more complicated than when contending with terrestrial species. Scavenging shark species may sometimes be identified by faunal evidence as it is not uncommon for shark teeth to become lodged in bone (Anderson et al., 2003; Ihama et al., 2009; Işcan and McCabe, 1995; Stock et al., 2017). Where teeth are absent, characteristics of bite marks may be used to estimate the offending shark's size. Quantitative measures such as interdental distance and bite circumference may be used to construct a size range but are insufficient to positively identify the shark species (Lowry et al., 2009). For the purposes of this text, morphological characteristics are given for four representative species of shark of varying size classes (Table 5.28).

*Figure 5.48 Bull shark (*Carcharhinus leucas*). "Bull shark (*Carcharhinus leucas*)" by Sylke Rohrlach is licensed under CC BY-SA 2.0.*

Figure 5.49 Hammerhead shark (Sphyrnidae spp). "Hammerhead shark" is in the public domain.

Table 5.28 Morphological Characteristics of Common Shark Species

	Bull Shark (*C. leucas*)	Tiger Shark (*G. cuvier*)	Great Hammerhead (*S. mokarran*)	Great White (*C. carcharias*)
Tooth shape	Triangular teeth with serrated edges[a]			
Conservation status	Near threatened	Near threatened	Endangered	Vulnerable
Typical length (cm)[a]	157–230	226–350	234–300	350–500
*Interdental distance range (mm)[b]	7.7–18.3	3.9–25.8	Unavailable	10.8–48.9
*Bite circumference range (mm)[b]	172–443	97–540	201–583	180–990

*As defined in Lowry et al. (2009):
• Interdental distance: distance between tips of the most labial teeth of neighboring tooth rows.
• Bite circumference: circumference of the arc of the upper and lower jaw.
Conservation status courtesy of IUCN Red List of Threatened Species.
[a]*Compagno et al. (2005).*
[b]*Lowry et al. (2009).*

Soft-tissue defects produced by sharks have clear edges that are often scalloped or serrated, a result of the notched morphology of shark teeth (Byard et al., 2000; Ihama et al., 2009). In the torso, wounds are often crescent-shaped or rounded with substantial excision of underlying tissue (Byard et al., 2000; Ihama et al., 2009; Lowry et al., 2009; Rathbun and Rathbun, 1997). Skin over excised areas may be left intact (Ihama et al., 2009). Amputations of the appendages

Table 5.29 Soft Tissue Defects Produced by Sharks	
Description	**Reported Locations**
Defined edges, often scalloped or serrated[a,b]	Margins of soft tissue defects
Triangular or rectangular flaps of skin[b]	Overlying excised tissue
Amputation of appendages[b,c]	Major joints
Incised wounds, no abrasion[a,b]	Appendages, adjacent to amputations
Crescent-shaped or round lacerations with significant tissue excision[a,b,c,d]	Abdominal region
	Thoracic cavity
	Buttocks, thighs

[a]*Byard et al. (2000).*
[b]*Ihama et al. (2009).*
[c]*Rathbun and Rathbun (1997).*
[d]*Lowry et al. (2009).*

are also common, often with fairly clean edges and in association with incised wounds (Byard et al., 2000; Ihama et al., 2009; Rathbun and Rathbun, 1997) (Table 5.29).

In bone, sharks produce pits, punctures, scratches, and gouges. On long bone diaphyses, pits and punctures often occur in an arch, reflecting the curvature of the jaw (İşcan and McCabe, 1995; Lowry et al., 2009). Depending on the relative strength of the affected element versus the shark jaw, punctures may occur in association with compression fractures (Allaire et al., 2012; Stock et al., 2017). Scratches are common in cortical bone, often in association with bone shaving produced by characteristic notches along the cutting edge of the tooth, seen in many shark species (Allaire et al., 2012; Rathbun and Rathbun, 1997; Stock et al., 2017). Linear gouging, resembling sharp force trauma, is common along the diaphyses of the long bones (Allaire et al., 2012; Byard et al., 2000; Ihama et al., 2009; Stock et al., 2017). In scavenging by tiger or bull sharks, parallel series of gouges may spiral around the diaphysis, a product of their bite-and-spin feeding behaviors (İşcan and McCabe, 1995; Stock et al., 2017). A summary of bone modifications commonly produced by shark feeding can be found in Table 5.30.

One shark species of note is the tiger shark (*Galeocerdo cuvier*) (Fig. 5.50), which has been frequently identified as a scavenger in the literature and exhibits interesting effects on human remains. Tiger sharks occasionally employ a bite-and-spin feeding behavior, characterized by an initial bite to an extremity followed by a lateral spin.

Table 5.30 Bone Modification by Sharks

Damage Category	Traits	Reported Locations
Pits and punctures	Crescent-shaped lines of pits/punctures[a,b]	Long bone diaphyses
	V-shaped punctures[c,d]	Long bone epiphyses
	Punctures without associated fractures[c,d]	Innominates
	Punctures with associated compression fractures[c,d]	Ribs
		Innominates
Scratches	Shallow, overlapping striations with bone shaving[c,d,e]	Clavicles
		Ribs
		Long bone diaphyses
Gouges	Incised linear gouges[b,c,d,f,g]	Long bone diaphyses
	Spiral gouges[a,d]	Long bone diaphyses

[a] Işcan and McCabe (1995).
[b] Lowry et al. (2009).
[c] Allaire et al. (2012).
[d] Stock et al. (2017).
[e] Rathbun and Rathbun (1997).
[f] Byard et al. (2000).
[g] Ihama et al. (2009).

*Figure 5.50 Tiger shark (*Galeocerdo cuvier*). "Tiger shark" by N. Hammerschlag, Oregon State University is licensed under CC BY-SA 2.0.*

The teeth work like a saw to strip soft tissue from the bone, and the behavior may leave the spiral gouges previously described (Işcan and McCabe, 1995). Recently, bull sharks (*Carcharhinus leucas*) have been observed to engage in a similar behavior (Stock et al., 2017). The tiger shark's digestive system is also unique. It is not uncommon for large chunks of food to be swallowed whole. Once in the stomach, food can be stored without digestion for a period of several weeks (Rathbun and Rathbun, 1997). Without digestion, remains will continue to decompose, although little is known about how the environment of the shark's digestive tract influences the decomposition rate (Rathbun and Rathbun, 1997).

Bodies recovered from the sea present unique challenges for recovery and identification of human remains. First, battering by waves and abrasion by sand can significantly modify skeletal evidence of trauma or postmortem damage (Stock et al., 2017). Bodies are also subject to drift, with some studies reporting body movement of over 500 km from the location where the individual entered the water, leaving investigators with an impossibly large pool of potential decedents (Giertsen and Morild, 1989; Kringsholm et al., 2001). Individual body parts may disarticulate and drift independently, even prior to scavenger involvement, and some sharks make significant seasonal migrations that could prevent recovery if portions of the body have been swallowed (Haglund, 1993). The Great White shark (*Carcharodon carcharias*) (Fig. 5.51), for example, has been observed to migrate over 2000 km (Weng et al., 2007). Although shark-inflicted damage is easily identifiable, interpretation may be skewed by other taphonomic forces in the aquatic environment. Further, the large ranges of shark species and

*Figure 5.51 Great white shark (*Carcharodon carcharias*). "Great white shark" is in the public domain.*

the potential for shark migration or bodily drift poses significant challenges to defining search areas when attempting to recover remains from the water.

Distinguishing Between Shark Attacks and Scavenging

Sharks are perhaps the most intensively studied aquatic scavengers, likely due to the media-perpetuated fear of shark attacks (Stock et al., 2017). Fortunately, shark scavenging and shark attacks are often easily differentiated. The majority of shark attacks are thought to be triggered by motivations other than hunger; some researchers have suggested that humans are unappetizing to most shark species due to our low proportions of body fat (Klimley, 1994). Most shark attacks (approximately 80%) are cases of mistaken identity, characterized as "hit-and-run" attacks because the sharks often abandon the victim after the initial encounter (Stock et al., 2017). Victims of hit-and-run attacks sustain only superficial abrasions from the shark's denticles (rough scales that cover the skin) and one or two bites with minimal tissue loss, with damage concentrated on appendages dangling in the water (Byard et al., 2000; Stock et al., 2017). Shark scavenging would not be expected to follow such patterns and would likely result in more significant damage. However, damage produced by intentional attacks—characterized as "bump-and-bite", "sneak", or provoked attacks—may be more difficult to distinguish from scavenging events (Stock et al., 2017).

Shark Summary

- Various species of shark found along the North American coastline; tiger sharks and bull sharks most frequently suspected of scavenging activity in the forensic literature
- Most shark attacks are "hit-and-run," characterized by only one or two bites, whereas shark scavenging will produce more extensive damage
- Amputation of limbs or significant excision of tissue, especially in the gluteal and thigh area, is common
- Parallel linear or spiral gouging on long bone diaphyses are diagnostic; punctures are also common and may fall in an arch that reflects jaw morphology
- Recovery of remains may be tricky; sharks are highly mobile and often swallow large portions of tissue and/or bone, while surviving body parts are subject to drift

FAMILY SUIDAE—SUIDS—WILD BOAR AND DOMESTIC PIG

Suids, including domestic pigs (*Sus scrofa domesticus*) and the introduced European wild boar (*Sus scrofa*), are extremely omnivorous with voracious appetites (Figs. 5.52–5.54). Habitat, behavior, and morphological characteristics of these species can be found in Tables 5.31 and 5.32. Wild boar incorporate small (~6%) but significant portions of animal matter into their diet (Taylor and Uvalde, 1999). Domestic pigs will famously consume almost anything put in

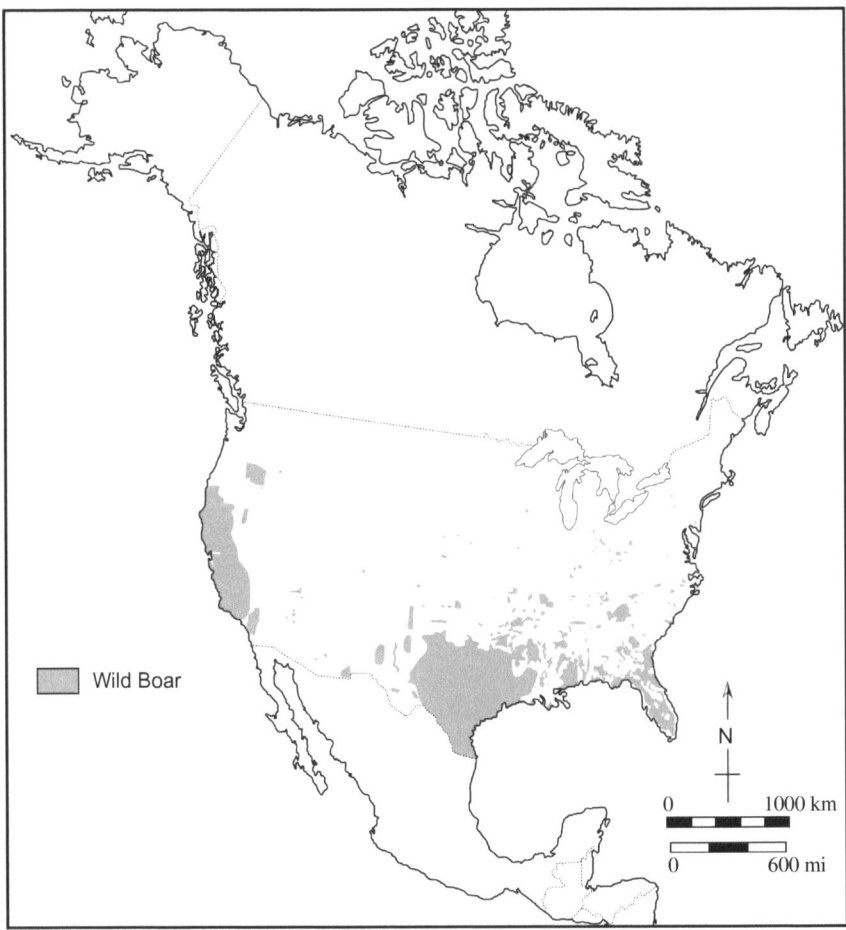

*Figure 5.52 Distribution of wild boar (*S. scrofa), *a species introduced from Europe, in North America. Domestic pigs (*S. scrofa domesticus), *not depicted, are found on farms throughout the continent, and occasionally, as pets. Distribution maps are meant only for general informational use, as species distributions change over time and individuals may occasionally be found outside of the depicted boundaries.* Distribution information compiled from Reid, F.A., 2006. The Princeton Field Guide to Mammals of North America. Houghton Mifflin Harcourt, Boston; Mississippi State University Extension. 2017. Wild pig info. Available from: http://wildpiginfo.msstate. edu/history-wild-pigs.html (accessed 31.07.17) (Mississippi State University Extension, 2017).

*Figure 5.53 Domestic pig (*Sus scrofa domesticus*). "Domestic pig" is in the public domain.*

*Figure 5.54 Wild boar (*Sus scrofa*). "Wild boar" is in the public domain.*

front of them. Both species are efficient, if perhaps surprising, scavengers. Canadian serial killer Robert "Willie" Pickton is believed to have utilized this efficiency, supposedly disposing of many of his victims by dismembering and feeding them to his domestic pigs, although no identifiable human remains were recovered within the pen (Cameron, 2011). Despite their efficiency at consuming tissue and destroying bone, relatively little research has been conducted on the taphonomic effects of scavenging by domestic pigs or wild boar.

Table 5.31 North American Suids*

Species Name	Conservation Status	Habitat
Wild boar (*S. scrofa*)	Least concern	Found in forested or dry brush areas. Occur throughout North America with major populations in California, Texas, and the southeastern United States.
Domestic pig (*S. scrofa domesticus*)		Raised as livestock or pets throughout North America.

Conservation status courtesy of IUCN Red List of Threatened Species.
** The collard peccary, or "javelina," (*Pecari tajacu*) is excluded from this volume as it is predominantly herbivorous (Corn & Warren, 1985).*
Habitat information adapted from Reid, F.A., 2006. The Princeton Field Guide to Mammals of North America. Houghton Mifflin Harcourt, Boston.

Table 5.32 Morphology and Behavior of Wild Boar

	Wild Boar
Peak daily activity[a]	Diurnal in winter; nocturnal or crepuscular in summer
Fur[a]	Coarse, black or brown fur
Weight range (lbs)[a]	110–330
Dental formula[b]	$\frac{3}{3}I\frac{1}{1}C\frac{4}{4}P\frac{3}{3}M$
Notable characteristics[a]	Pronounced, curving canines (tusks)

[a] Reid (2006).
[b] Adams and Crabtree (2011).

The thoracic cavity is believed to be the primary scavenging target of swine, based on Galdikas's (1978) observation of a Bornean bearded pig scavenging an orangutan carcass. This is further supported by a case study of pig scavenging on human remains (Berryman, 2001). Although most bones in this case were recovered in relative anatomical position, Berryman (2001) reported extensive damage to the midline of the body, which he attributed to attempts to access the viscera. The damage included fractures to the medial clavicles, crushing of the sternal ribs and fractures to the vertebral ends, and destruction of the pubic bones; there was also considerable damage to the facial bones and bones of the hands (Berryman, 2001).

Greenfield (1988) observed that swine have a preference for bones which could be easily picked up and chewed, resulting in the near-complete destruction of lower density, small, and mid-sized elements. Domínguez-Solera and Domínguez-Rodrigo (2009) also noted that scavenging by swine results in extensive fragmentation, if not complete destruction, of smaller elements, including components of the axial skeleton. Long bone epiphyses are also vulnerable to destruction (Greenfield, 1988). Bones that are too large to be picked up are rolled to access tissue on all sides, but tend to remain fairly close to their original location. Long scores with flat bottoms predominate on surviving compact bone, a consequence of suids scraping tissue from bone with their mandibular incisors (Berryman, 2001; Domínguez-Solera and Domínguez-Rodrigo, 2009; Greenfield, 1988). Domínguez-Solera and Domínguez-Rodrigo (2009) also reported L-shaped punctures in surviving long bone epiphyses, reflecting pig premolar morphology (Fig. 5.55). See Table 5.33 for a summary of bone modifications commonly produced by suid scavenging.

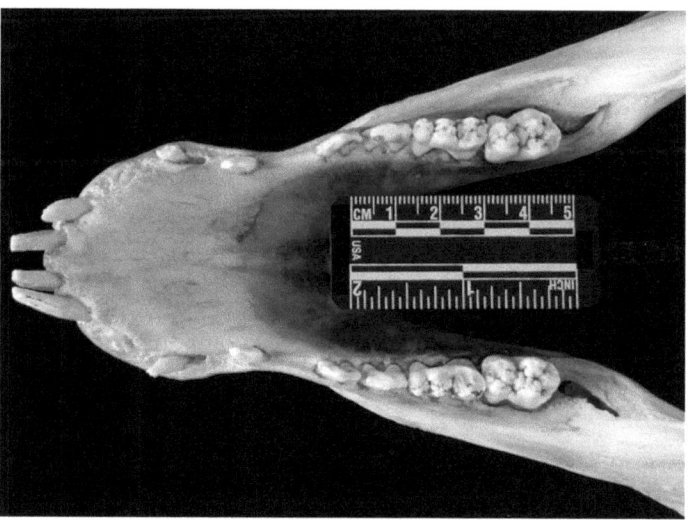

*Figure 5.55 Domestic pig (*S. scrofa domesticus*) dentition. Image depicts a juvenile individual; note distinctive morphology of the posterior teeth (all deciduous premolars).* Photograph by David Tuttle.

Table 5.33 Bone Modification by Suids

Damage Category	Traits	Reported Locations
Tooth marks	"L"-shaped punctures[a]	Long bone epiphyses
	Long, broad, shallow scores[a,c]	Flat bones
		Proximal and distal portions of diaphyses
	Shovel-shaped, flat furrows[a,b]	Long bone epiphyses
Breakage	Crushing[c]	Long bone epiphyses
	Fractures[c]	Medial clavicle
		Sternum
		Sternal, vertebral rib ends
	Destruction of features[b]	Vertebral processes
		Iliac tubercle
		Long bone epiphyses
	Destruction of elements[a,b]	Vertebrae
		Long bones
		Innominate

[a]Dominguez-Solera and Dominguez-Rodrigo (2009).
[b]Greenfield (1988).
[c]Berryman (2001).

Suid Summary

• Two relevant species: domestic pigs and European wild boar
• Domestic pig distribution ubiquitous throughout the continent; wild boar predominant in California, Texas, and the southeastern United States, with sporadic populations in other parts of the United States
• Both species omnivorous and infamous for appetites that include all types of food (and nonfood) items
• Evidence suggests that suids preferentially destroy the thoracic cavity to access internal organs, and will chew small bones to their destruction
• Tooth marks on bone include shallow scores, L-shaped punctures, and shovel-shaped flat furrows
• Bone breakage can be extensive and includes fractures and crushing damage on a variety of elements and features.

FAMILY URSIDAE—URSIDS—BEARS

The distribution of bear species in North America is presented in Fig. 5.56, while habitat preferences, behavior, and morphological

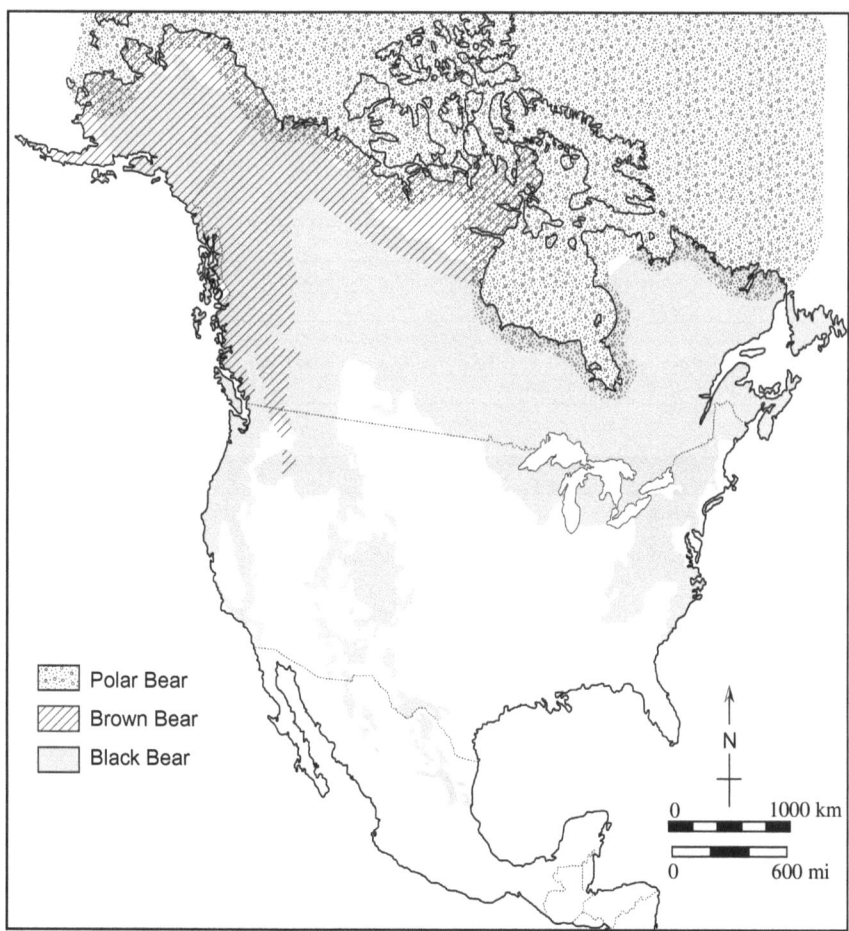

Figure 5.56 Distribution of American black bears (U. americanus), brown bears (U. arctos), and polar bears (U. maritimus) in North America. Note that polar bears are depicted throughout the Arctic, as their range includes Arctic sea ice. Distribution maps are meant only for general informational use, as species distributions change over time and individuals may occasionally be found outside the depicted boundaries. Distribution information compiled from Reid, F.A., 2006. The Princeton Field Guide to Mammals of North America. Houghton Mifflin Harcourt, Boston; IUCN 2017. The IUCN Red List of Threatened Species. Version 2017-1. Available from: http://www.iucnredlist.org/ (accessed 31.07.17).

descriptions are provided in Tables 5.34 and 5.35. Bears are omnivorous, with mixed diets including vegetation as well as meat in the form of fish (salmon in particular), ungulates, or—in the case of the polar bear (*Ursus maritimus*)—marine mammals (Mowat and Heard, 2006; Thiemann et al., 2008; Zager and Beecham, 2006) (Fig. 5.57). Bears are opportunistic feeders as well, and will scavenge when the opportunity presents itself. Because of their large size, they are

Table 5.34 North American Bear Species

Species Name	Conservation Status	Habitat
American black bear (*U. americanus*)	Least concern	Occupy forests and swamps in the east, prefer mountains or tundra in the west.
Brown bear (*U. arctos*)	Least concern	May occupy tundra, mountains, coastal forests, and open countryside near rivers.
Polar bear (*U. maritimus*)	Vulnerable	Tundra, often along rocky coasts.

Conservation status courtesy of IUCN Red List of Threatened Species.
Habitat information adapted from Reid, F.A., 2006. The Princeton Field Guide to Mammals of North America. Houghton Mifflin Harcourt, Boston.

Table 5.35 Morphological and Behavioral Characteristics of Bears

	Black Bear	Brown Bear	Polar Bear
Peak daily activity[a]	Diurnal	Crepuscular	Crepuscular
Sociality[a]	Solitary, except when feeding	Solitary, small family groups	Solitary, except females with cubs
Fur[a]	Black or very dark brown	Warm brown, sometimes blonde	White, sometimes yellowed
Weight range (lbs)[a]	100–900	180–2100	330–1600
Dental formula[b]	$\frac{3}{3}I\ \frac{1}{1}C\ \frac{2-4}{2-4}P\ \frac{2}{3}M$		

[a] *Reid (2006).*
[b] *Adams and Crabtree (2011).*

capable of producing extensive damage to human remains. However, most frequently, scavenging by bears produces only slight to moderate modification of the skeleton (Saladié et al., 2013). In long bones, damage is often concentrated at epiphyses but is much less extensive than the marks typically produced by prolonged canid gnawing (Bright, 2011; Carson et al., 2000).

Like that of other carnivores, bone modification by bears is characterized by pits, punctures, scores, and furrows (Table 5.36). The canine teeth of bears are extremely prominent, but pits and punctures seen in bear-scavenged cases are often smaller than would be expected given both the size of the canines and the genus's incredible jaw strength. Bright (2011) has suggested that these undersized pits or punctures may instead be formed by the raised cusps of the carnassial teeth

*Figure 5.57 Polar bear (*Ursus maritimus*). "Polar bear" by G. Gellinger is in the public domain.*

Table 5.36 Bone Modification by Bears		
Damage Category	**Traits**	**Reported Locations**
Margins	Crenulations[a]	Innominates (ilia, ischia) Vertebral margins of scapulae
	Crushing[a]	Cortical bone adjacent to tooth marks
	Scalloping[b]	Margins of epiphyses
	Splintering[b]	Vertebrae
		Ribs
Tooth marks	Pits[b]	Cranium
	Punctures[b]	Scapulae Vertebrae
	Scores[a]	Long bone epiphyses
	Furrows[a]	Innominates
Breakage	Spiral, radial fractures[b,c]	Long bone diaphyses
	Obliteration of features[b,c]	Mandibular processes Vertebral processes
	Destruction of elements[b,c]	Sternum Vertebrae

[a] *Saladié et al. (2013).*
[b] *Carson et al. (2000).*
[c] *Murad and Boddy (1987).*

(the upper fourth premolar and lower first molar, adapted for shearing meat) (Fig. 4.8). Scores created by bear scavenging have a U-shaped cross section and flat bottoms, reflecting the bunodont molar morphology of bears; in other words, their molars have rounded cusps adapted to their omnivorous diet, a trait they share with humans (Saladié et al., 2013). These scores occur in clusters as either (1) disordered and superficial or (2) deep, wide, and parallel (Saladié et al., 2013). In addition, tooth marks produced by bear scavenging are often associated with large areas of crushed cortical bone (Saladié et al., 2013).

In terms of bodily destruction, preliminary research suggests that bears preferentially consume the organs within the thoracic cavity (e.g., the heart and lungs), followed by destruction of the sternum and ribs to access the shoulder girdle, while lower extremities are exploited last (Carson et al., 2000). This consumption pattern results in preferential destruction of the axial skeleton, especially the sternum, vertebrae, and ribs. Heavy destruction of the axial skeleton may be a pattern that can help distinguish scavenging by bears from that of other large-bodied carnivores. However, other research suggests that the pattern is not entirely consistent, although there is a proposed ecological explanation. In a study conducted by Bright (2011), the pattern of consumption deviated from the identified pattern in the three bear-scavenged cases described by Carson et al. (2000). While Carson's work was conducted on forensic casework recovered in spring or summer in the American southwest, Bright's thesis research was conducted on five pig carcasses deposited in northern California in early November (Bright, 2011; Carson et al., 2000).

To explain the discrepancy, Bright (2011) suggests that the consumption pattern of bears may shift before hibernation, a period of fasting in which bear activity is significantly reduced and which coincides with the genus's reproductive season. North American black and brown bears. begin preparing for hibernation in the fall, accumulating energy reserves to meet the energetic demands of the winter season (Figs. 5.58 and 5.59). During this period, referred to as hyperphagia, bears may double their daily meat intake to increase their body mass primarily by building up their deposits of body fat (Hilderbrand et al., 1999; Nelson et al., 1983). Bright (2011) observed scavenging by the American black bear during this hyperphagic period and found that the bears preferentially consumed fatty skin, followed by the shoulder and haunch, before consuming the protein-rich thoracic organs. This

*Figure 5.58 American black bears (*Ursus americanus*). Photograph by R.T. Kramer.

*Figure 5.59 Brown bear (*Ursus arctos*). Photograph by R.T. Kramer.

is consistent with prior research that demonstrates how a diet rich in fat, which is more easily converted to body fat than protein, promotes the biochemical metabolic processes which sustain a bear during hibernation (Nelson et al., 1983). It should be noted that polar bears experience hibernation differently than their more southern relatives, engaging in relatively active fasting during warmer summer months when their access to seals (their primary prey) is restricted due to retreating sea ice (Atkinson and Ramsay, 1995; Nelson et al., 1983).

The skull and forelimbs are usually disarticulated early on in cases of bear scavenging, but given their large size, bears are able to easily transport entire carcasses. Bears tend to scavenge at higher elevations and have been shown to drag carcasses over 130 m upslope (Bright, 2011). Temperature at higher elevations tends to be cooler, which reduces microbial and insect activity. However, documented elevation changes are relatively small and would likely not have a significant impact on spoilage rate (Bright, 2011). It may also be possible that there are fewer competitive species present at higher elevations, which has been shown to influence bear utilization of wolf-killed carcasses in Wyoming (Green et al., 1997).

Bears also preferentially deposit remains in areas with high horizontal cover, such as in thickets of dense vegetation or at the base of fallen trees, likely to prevent visual detection by competing scavengers (Bright, 2011; Cristescu et al., 2014). European brown bear species have also been documented caching a carcass by scraping local debris over it, creating a roughly circular covering (Elgmork, 1982). Although North American brown bear counterparts in Canada have been documented caching ungulates, the specific nature of the identified caching sites has not been explicitly described, so it is unclear whether this type of cache is utilized by North American bears (Cristescu et al., 2014).

Ursid Summary
- Three North American species: brown (or grizzly), American black, and polar bears
- Depending on species, distribution ranges across Canada and Alaska (all bears) as well as different regions throughout the United States (black bears)
- Generally omnivorous but opportunistic scavengers
- Scavenging damage to skeleton typically slight despite bears' size, concentrated at epiphyses
- Tooth marks include pits, punctures, scoring, and furrows associated with large areas of crushed cortical bone
- Pits and punctures often smaller than those typically associated with canid scavenging

- Skull and forelimbs often disarticulated from remains; but entire body can be transported as a unit
- Axial skeleton often damaged due to targeted feeding on the thoracic cavity; however, this pattern is altered during the fall season while bears prepare to hibernate

REFERENCES

OVERALL

Adams, B., Crabtree, P., 2011. Comparative Osteology: A Laboratory and Field Guide of Common North American Animals. Elsevier Science, Burlington.

Cornell Lab of Ornithology, 2015. All About Birds. Cornell University. Available from: https://www.allaboutbirds.org/. (accessed 31.07.17).

IUCN 2017. The IUCN Red List of Threatened Species. Version 2017-1. Available from: http://www.iucnredlist.org/. (accessed 31.07.17).

Reid, F.A., 2006. The Princeton Field Guide to Mammals of North America. Houghton Mifflin Harcourt, Boston.

Vander Wall, S.B., 1990. Food-hoarding mammals. In: Vander Wall, S.B. (Ed.), Food Hoarding in Animals. University of Chicago Press, Chicago, pp. 217−282.

CANIDS—FAMILY CANIDAE—DOGS, COYOTES, FOXES, AND WOLVES

Anandarup, B., Bhattacharjee, D., Paul, M., Singh, A., Gade, P.R., Shrestha, P., et al., 2016. The meat of the matter: a rule of thumb for scavenging dogs? Ethol. Ecol. Evol. 28 (4), 427−440.

Bankaitis, J., 2012. Examination of scavenging associated with wolves. Unpublished Masters thesis. Department of Anthropology. The University of Montana, Missoula.

Binford, L.R., 1981. Patterns of bone modifications produced by nonhuman agents. In: Binford, L.R. (Ed.), Bones: Ancient Men and Modern Myths. Academic Press, San Diego, pp. 35−86.

Colard, T., Delannoy, Y., Naji, S., Gosset, D., Hartnett, K., Bécart, A., 2015. Specific patterns of canine scavenging in indoor settings. J. Forensic Sci. 60 (2), 495−500.

Fonseca, G.M., Mora, E., Lucena, J., Cantin, M., 2015. Forensic studies of dog attacks on humans: A focus on bite mark analysis. Res. Rep. Forensic Med. Sci. 5, 39−51.

Fonseca, G.M., Palacios, R., 2013. An unusual case of predation: dog pack or cougar attack? J. Forensic Sci. 58 (1), 224−227.

Haglund, W.D., 1997a. Dogs and coyotes: postmortem involvement with human remains. In: Haglund, W.D., Sorg, M.H. (Eds.), Forensic Taphonomy: The Postmortem Fate of Human Remains. CRC Press, Boca Raton, pp. 367−382.

Haynes, G., 1982. Utilization and skeletal disturbances of North American prey carcasses. Arctic 35 (2), 266−281.

Hernández-Carrasco, M., Pisani, J.M.A., Scarso-Giaconi, F., Fonseca, G.M., 2016. Indoor postmortem mutilation by dogs: confusion, contradictions, and needs from the perspective of the forensic veterinarian medicine. J. Vet. Behav. Clin. Appl. Res. 15, 56−60.

Kjorlien, Y.P., Beattie, O.B., Peterson, A.E., 2009. Scavenging activity can produce predictable patterns in surface skeletal remains scattering: observations and comments from two experiments. Forensic Sci. Int. 188, 103−106.

Metz, M.C., Smith, D.W., Vucetich, J.A., Stahler, D.R., Peterson, R.O., 2012. Seasonal patterns of predation for gray wolves in the multi-prey system of Yellowstone National Park. J. Anim. Ecol. 81 (3), 553–563.

Nelson, M.E., 2011. Killing and caching of an adult white-tailed deer, *Odocoileus virginianus*, by a single gray wolf, *Canis lupus*. Can. Field Nat. 125 (2), 162–164.

Pokines, J., 2014. Faunal dispersal, reconcentration, and gnawing damage to bone in terrestrial environments. In: Pokines, J., Symes, S. (Eds.), Manual of Forensic Taphonomy. CRC Press, Boca Raton, pp. 201–248.

Rothschild, M.A., Schneider, V., 1997. On the temporal onset of postmortem animal scavenging:"Motivation" of the animal. Forensic Sci. Int. 89 (1), 57–64.

Santoro, V., Smaldone, G., Lozito, P., Smaldone, M., Introna, F., 2011. A forensic approach to fatal dog attacks. A case study and review of the literature. Forensic Sci. Int. 206 (1), e37–e42.

Steadman, D.W., Worne, H., 2007. Canine scavenging of human remains in an indoor setting. Forensic Sci. Int. 173 (1), 78–82.

Tsokos, M., Schulz, F., 1999. Indoor postmortem animal interference by carnivores and rodents: report of two cases and review of the literature. Int. J. Leg. Med. 112 (2), 115–119.

Willey, P., Snyder, L., 1989. Canid modification of human remains: implications for time-since-death estimations. J. Forensic Sci. 34 (4), 894–901.

Young, A., Márquez-Grant, N., Stillman, R., Smith, M.J., Korstjens, A.H., 2015. An investigation of red fox (*Vulpes vulpes*) and Eurasian badger (*Meles meles*) scavenging, scattering, and removal of deer remains: forensic implications and applications. J. Forensic Sci. 60 (S1), S39–S55.

FAMILY CATHARTIDAE—NEW WORLD VULTURES

Ballejo, F., Fernández, F.J., De Santis, L.J.M., Montalvo, C.I., 2016. Taphonomy and dispersion of bones scavenged by New World vultures and caracaras in Northwestern Patagonia: implications for the formation of archaeological sites. Archaeol. Anthropol. Sci. 8 (2), 305–315.

Beck, J., Ostericher, I., Sollish, G., De Leon, J., 2015. Animal scavenging and scattering and the implications for documenting the deaths of undocumented border crossers in the Sonoran Desert. J. Forensic Sci. 60 (S1), S11–S20.

Buckley, N.J., 1996. Food finding and the influence of information, local enhancement, and communal roosting on foraging success of North American vultures. Auk 113 (2), 473–488.

Coleman, J.S., Fraser, J.D., 1987. Food habits of black and turkey vultures in Pennsylvania and Maryland. J. Wildl. Manage. 51 (4), 733.

Collins, P.W., Noel, F.R.S., Emslie, S.D., 2000. Faunal remains in California condor nest caves. Condor 102 (1), 222–227.

Dabbs, G.R., Martin, D.C., 2013. Geographic variation in the taphonomic effect of vulture scavenging: the case for southern Illinois. J. Forensic Sci. 58 (S1), S20–S25.

Domínguez-Solera, S., Domínguez-Rodrigo, M., 2011. A taphonomic study of a carcass consumed by griffon vultures (*Gyps fulvus*) and its relevance for the interpretation of bone surface modifications. Archaeol. Anthropol. Sci. 3 (4), 385–392.

Kiff, L.F., 2000. The current status of North American vultures. In: Chancellor, R.D., Meyburg, B. (Eds.), Raptors at Risk: Proceedings of the 5th World Conference on Birds of Prey and Owls. Hancock House, Towcester, pp. 175–189.

Klein, A.A., 2013. Vulture scavenging of pig remains at varying grave depths. Unpublished Masters thesis. Department of Anthropology. Texas State University, San Marcos.

Platt, S.G., Gukian, T., Meraz, R.E., Ritzi, C.M., Rainwater, T.R., 2015. Exhumation of buried mammal carrion by turkey vultures. J. Raptor Res. 49 (4), 518–520.

Prior, K.A., Weatherhead, P.J., 1991. Competition at the carcass: opportunities for social foraging by turkey vultures in southern Ontario. Can. J. Zool. 69 (6), 1550–1556.

Reeves, N.M., 2009. Taphonomic effects of vulture scavenging. J. Forensic Sci. 54 (3), 523–528.

Smith, H.R., DeGraaf, R.M., Miller, R.S., 2002. Exhumation of food by turkey vulture. J. Raptor Res. 36 (2), 144–145.

Spradley, M.K., Hamilton, M.D., Giordano, A., 2012. Spatial patterning of vulture scavenged human remains. Forensic Sci. Int. 219, 57–63.

CERVIDS—FAMILY CERVIDAE—DEER

Brothwell, D., 1976. Further evidence of bone chewing by ungulates: the sheep of North Ronaldsay, Orkney. J. Archaeol. Sci. 3 (2), 179–182.

Cáceres, I., Esteban-Nadal, M., Bennàsar, M., Fernández-Jalvo, Y., 2011. Was it the deer or the fox? J. Archaeol. Sci. 38 (10), 2767–2774.

Gambín, P., Ceacero, F., Garcia, A.J., Landete-Castillejos, T., Gallego, L., 2017. Patterns of antler consumption reveal osteophagia as a natural mineral resource in key periods for red deer (Cervus elaphus). Eur. J. Wildl. Res. 63 (2), 39.

Hutson, J.M., Burke, C.C., Haynes, G., 2013. Osteophagia and bone modifications by giraffe and other large ungulates. J. Archaeol. Sci. 40 (12), 4139–4149.

Johnson, D., Haynes, C., 1985. Camels as taphonomic agents. Quater. Res. 24 (3), 365–366.

Keating, K., 1990. Bone chewing by Rocky Mountain bighorn sheep. Great Basin Nat. 50 (1), 89.

Kierdorf, U., 1994. A further example of long-bone damage due to chewing by deer. Int. J. Osteoarchaeol. 4 (3), 209–213.

Meckel, L.A., McDaneld, C.P., Wescott, D.J., 2017. White-tailed deer as a taphonomic agent: photographic evidence of white-tailed deer gnawing on human bone. J. Forensic Sci. Available from: http://dx.doi.org/10.1111/1556-4029.13514.

Sutcliffe, A.J., 1973. Similarity of bones and antlers gnawed by deer to human artefacts. Nature 246, 428–430.

FAMILY CORVIDAE—CORVIDS—MAGPIES, CROWS, AND RAVENS

Allen, M.L., Taylor, A.P., 2013. First record of scavenging by a Western screech-owl (Megascops kennicottii). Wilson J. Ornithol. 125 (2), 417–419.

Asamura, H., Fukushima, H., Kobayashi, K., Takayanagi, K., Ota, M., 2003. Unusual characteristic patterns of postmortem injuries. J. Forensic Sci. 49 (3), 592–594.

Ballejo, F., Fernández, F.J., De Santis, L.J.M., Montalvo, C.I., 2016. Taphonomy and dispersion of bones scavenged by New World vultures and caracaras in Northwestern Patagonia: implications for the formation of archaeological sites. Archaeol. Anthropol. Sci. 8 (2), 305–315.

Bauer, J.W., Logan, K.A., Sweanor, L.L., Boyce, W.M., 2005. Scavenging behavior in puma. Southw. Nat. 50 (4), 466–471.

de Kort, S.R., Clayton, N.S., 2006. An evolutionary perspective on caching by corvids. Proc. R. Soc. B Biol. Sci. 273 (1585), 417–423.

Heinrich, B., 1988. Winter foraging at carcasses by three sympatric corvids, with emphasis on recruitment by the raven, Corvus corax. Behav. Ecol. Sociobiol. 23 (3), 141–156.

Kane, A., Jackson, A.L., Ogada, D.L., Monadjem, A., McNally, L., 2014. Vultures acquire information on carcass location from scavenging eagles. Proc. R. Soc. B Biol. Sci. Available from: http://dx.doi.org/10.1098/rspb.2014.1072.

Komar, D., Beattie, O., 1998. Identifying bird scavenging in fleshed and dry remains. J. Can. Soc. Forensic Sci. 31 (3), 177–188.

Morton, R.J., Lord, W.D., 2006. Taphonomy of child-sized remains: a study of scattering and scavenging in Virginia, USA. J. Forensic Sci. 51 (3), 475–479.

Roll, P., Rous, F., 1991. Injuries by chicken bills: characteristic wound morphology. Forensic Sci. Int. 52 (1), 25–30.

Stahler, D., Heinrich, B., Smith, D., 2002. Common ravens, *Corvus corax*, preferentially associate with grey wolves, *Canis lupus*, as a foraging strategy in winter. Anim. Behav. 64 (2), 283–290.

Young, A., Stillman, R., Smith, M.J., Korstjens, A.H., 2014. An experimental study of vertebrate scavenging behavior in a northwest European woodland context. J. Forensic Sci. 59 (5), 1333–1342.

ORDER CROCODILIA—CROCODILIANS

Bangs, P.M., 2014. Decomposition at three aquatic and terrestrial sites in southern Louisiana. Unpublished Masters thesis. Department of Geography and Anthropology. Louisiana State University, Baton Rouge.

Bartlett, R.D., Bartlett, P.P., 2011. Florida's Turtles, Lizards, and Crocodilians. University Press of Florida, Gainesville.

Cleuren, J., de Vree, F., 1992. Kinematics of the jaw and hyolingual apparatus during feeding in *Caiman crocodilus*. J. Morphol. 212 (2), 141–154.

Delany, M.F., Abercrombie, C.L., 1986. American alligator food habits in northcentral Florida. J. Wildl. Manage. 50 (2), 348–353.

Drumheller, S.K., Brochu, C.A., 2014. A diagnosis of *Alligator mississippiensis* bite marks with comparisons to existing crocodylian datasets. Ichnos 21 (2), 131–146.

Drumheller, S.K., Brochu, C.A., 2016. Phylogenetic taphonomy: a statistical and phylogenetic approach for exploring taphonomic patterns in the fossil record using crocodylians. Palaios 31 (10), 463–478.

Haddad, V., Fonseca, W.C., 2011. A fatal attack on a child by a black caiman (*Melanosuchus niger*). Wilder. Environ. Med. 22 (1), 62–64.

Harding, B.E., Wolf, B.C., 2006. Alligator attacks in southwest Florida. J. Forensic Sci. 51 (3), 674–677.

Joanen, T., 1969. Nesting ecology of alligators in Louisiana. Proc. Ann. Conf. Southe. Assoc. Game Fish Commis. 23, 141–151.

Langley, R.L., 2005. Alligator attacks on humans in the United States. Wilder. Environ. Med. 16 (3), 119–124.

Langley, R.L., 2010. Adverse encounters with alligators in the United States: an update. Wilder. Environ. Med. 21 (2), 156–163.

McNease, L., Joanen, T., 1981. Nutrition of alligators. Paper presented at: 1st Annual Alligator Production Conference. University of Florida.

Nifong, J.C., 2016. Living on the edge: trophic ecology of *Alligator mississippiensis* (American alligator) with access to a shallow estuarine impoundment. Bull. Florida Mus. Nat. Hist. 54 (2), 13–49.

Njau, J.K., Blumenschine, R.J., 2006. A diagnosis of crocodile feeding traces on larger mammal bone, with fossil examples from the Plio-Pleistocene Olduvai Basin, Tanzania. J. Hum. Evol. 50 (2), 142–162.

Platt, S.G., Thorbjarnarson, J.B., Rainwater, T.R., Martin, D.R., 2013. Diet of the American crocodile (*Crocodylus acutus*) in marine environments of coastal Belize. J. Herpetol. 47 (1), 1–10.

Shepherd, S.M., Shoff, W.H., 2014. An urban northeastern United States alligator bite. Am. J. Emerg. Med. 32 (5), 487.e481-487.e483.

Sinton, T.J., Byard, R.W., 2016. Pathological features of fatal crocodile attacks in northern Australia, 2005–2014. J. Forensic Sci. 61 (6), 1553–1555.

Wheatley, P.V., Peckham, H., Newsome, S.D., Koch, P.L., 2012. Estimating marine resource use by the American crocodile *Crocodylus acutus* in southern Florida, USA. Mar. Ecol. Progr. Series 447, 211–229.

Wolf, B.C., Harding, B.E., 2014. Fatalities due to indigenous and exotic species in Florida. J. Forensic Sci. 59 (1), 155–160.

FAMILY DIDELPHIDAE—*DIDELPHIS VIRGINIANA*—VIRGINIA OPOSSUM

DeVault, T.L., Olson, Z.H., Beasley, J.C., Rhodes Jr, O.E., 2011. Mesopredators dominate competition for carrion in an agricultural landscape. Basic Appl. Ecol. 12 (3), 268–274.

Jeong, Y., Jantz, L.M., Smith, J., 2016. Investigation into seasonal scavenging patterns of raccoons on human decomposition. J. Forensic Sci. 61 (2), 467–471.

King, K.A., Lord, W.D., Ketchum, H.R., O'Brien, R.C., 2016. Postmortem scavenging by the Virginia opossum (*Didelphis virginiana*): impact on taphonomic assemblages and progression. Forensic Sci. Int. 266, 576.e1–576.e6.

Megyesi, M.S., Nawrocki, S.P., Haskell, N.H., 2005. Using accumulated degree-days to estimate the postmortem interval from decomposed human remains. J. Forensic Sci. 50 (3), 618–626.

Morton, R.J., Lord, W.D., 2006. Taphonomy of child-sized remains: a study of scattering and scavenging in Virginia, USA. J. Forensic Sci. 51 (3), 475–479.

Olson, Z.H., Beasley, J.C., Rhodes, O.E., 2016. Carcass type affects local scavenger guilds more than habitat connectivity. PLoS One 11 (2), e0147798.

Synstelien, J.A., Klippel, W.E., 2005. Raccoon (*Procyon lotor*) foraging as a taphonomic agent of soft tissue modification and scene alteration. Proc. Am. Acad. Forensic Sci. 11, 333–334.

FAMILY FELIDAE—FELIDS—WILD AND DOMESTIC CATS

Álvarez, M.C., Kaufmann, C.A., Massigoge, A., Gutiérrez, M.A., Rafuse, D.J., Scheifler, N.A., et al., 2012. Bone modification and destruction patterns of Leporid carcasses by Geoffroy's cat (*Leopardus geoffroyi*): an experimental study. Quater. Int. 278, 71–80.

Bauer, J.W., Logan, K.A., Sweanor, L.L., Boyce, W.M., 2005. Scavenging behavior in puma. Southw. Nat. 50 (4), 466–471.

Bischoff-Mattson, Z., Mattson, D., 2009. Effects of simulated mountain lion caching on decomposition of ungulate carcasses. West. North Am. Nat. 69 (3), 343–350.

Bradshaw, J., Healey, L., Thorne, C., Macdonald, D., Arden-Clark, C., 2000. Differences in food preferences between individuals and populations of domestic cats *Felis silvestris catus*. Appl. Anim. Behav. Sci. 68 (3), 257–268.

Elbroch, L.M., Wittmer, H.U., 2013. Nuisance ecology: do scavenging condors exact foraging costs on pumas in Patagonia? PLoS One 8 (1), 1–8.

Kaufmann, C.A., Rafuse, D.J., González, M.E., Álvarez, M.C., Massigoge, A., Scheifler, N.A., et al., 2016. Carcass utilization and bone modifications on guanaco killed by puma in northern Patagonia, Argentina. Quatern. Int. Available from: https://doi.org/10.1016/j.quaint.2016.03.003.

Krofel, M., Kos, I., Jerina, K., 2012. The noble cats and the big bad scavengers: Effects of dominant scavengers on solitary predators. Behav. Ecol. Sociobiol. 66 (9), 1297–1304.

Moran, N., O'Connor, T., 1992. Bones that cats gnawed upon: a case study in bone modification. Circaea 9 (1), 27–34.

Mondini, M., Muñoz, A.S., 2008. Pumas as taphonomic agents: a comparative analysis of actualistic studies in the Neotropics. Quatern. Int. 180 (1), 52–62.

Pickering, T.R., 2001. Carnivore voiding: a taphonomic process with the potential for the deposition of forensic evidence. J. Forensic Sci. 46 (2), 406–411.

Pickering, T.R., Carlson, K.J., 2004. Baboon taphonomy and its relevance to the investigation of large felid involvement in human forensic cases. Forensic Sci. Int. 144 (1), 37–44.

Pokines, J., 2014. Faunal dispersal, reconcentration, and gnawing damage to bone in terrestrial environments. In: Pokines, J., Symes, S. (Eds.), Manual of Forensic Taphonomy. CRC Press, Boca Raton, pp. 201–248.

Rippley, A., Larison, N.C., Moss, K.E., Kelly, J.D., Bytheway, J.A., 2012. Scavenging behavior of *Lynx rufus* on human remains during the winter months of southeast Texas. J. Forensic Sci. 57 (3), 699–705.

Rossi, M., Shahrom, A., Chapman, R., Vanezis, P., 1994. Postmortem injuries by indoor pets. Am. J. Forensic Med. Pathol. 15 (2), 105–109.

FAMILY PROCYONIDAE—*PROCYON LOTOR*—NORTHERN RACCOON

Hannigan, A., 2015. A descriptive study of forensic implications of raccoon scavenging in Maine. Unpublished Honors thesis. Department of Anthropology. University of Maine, Orono.

Jeong, Y., Jantz, L.M., Smith, J., 2016. Investigation into seasonal scavenging patterns of raccoons on human decomposition. J. Forensic Sci. 61 (2), 467–471.

Komar, D., Beattie, O., 1998. Identifying bird scavenging in fleshed and dry remains. J. Can. Soc. Forensic Sci. 31 (3), 177–188.

Morton, R.J., Lord, W.D., 2006. Taphonomy of child-sized remains: a study of scattering and scavenging in Virginia, USA. J. Forensic Sci. 51 (3), 475–479.

Smith, J.K., 2015. Raccoon scavenging and the taphonomic effects on early human decomposition and PMI estimation. Unpublished Masters thesis. Department of Anthropology. University of Tennessee, Knoxville.

Steadman, D.W., Dautartas, A., Mundorff, A., Vidoli, G., Jantz, L., 2016. Differential raccoon scavenging among pig, rabbit, and human subjects. Proc. Am. Acad. Forensic Sci. 22, 192.

Synstelien, J.A., 2013. Raccoon modification of human skeletal remains. *American Journal of Physical Anthropology* 150(S56): 268. Program of the 82nd Annual Meeting of the American Association of Physical Anthropologists, Knoxville.

Synstelien, J.A., Klippel, W.E., 2005. Raccoon (*Procyon lotor*) foraging as a taphonomic agent of soft tissue modification and scene alteration. Proc. Am. Acad. Forensic Sci. 11, 333–334.

ORDER RODENTIA—RODENTS

Haglund, W.D., 1992. Contribution of rodents to postmortem artifacts of bone and soft tissue. J. Forensic Sci. 37 (6), 1459–1465.

Haglund, W.D., 1997b. Rodents and human remains. In: Haglund, W.D., Sorg, M.H. (Eds.), Forensic Taphonomy: The Postmortem Fate of Human Remains. CRC Press, Boca Raton, pp. 405–414.

Kay, E.H., Hoekstra, H.E., 2008. Rodents. Curr. Biol. 18 (10), R406–R410.

Klippel, W.E., Synstelien, J.A., 2007. Rodents as taphonomic agents: bone gnawing by brown rats and gray squirrels. J. Forensic Sci. 52 (4), 765–773.

Pokines, J.T., 2015. Taphonomic alterations by the rodent species woodland vole (*Microtus pinetorum*) upon human skeletal remains. Forensic Sci. Int. 257, e16—e19.

Ropohl, D., Scheithauer, R., Pollak, S., 1995. Postmortem injuries inflicted by domestic golden hamster: morphological aspects and evidence by DNA typing. Forensic Sci. Int. 72 (2), 81—90.

Tsokos, M., Matschke, J., Gehl, A., Koops, E., Püschel, K., 1999. Skin and soft tissue artifacts due to postmortem damage caused by rodents. Forensic Sci. Int. 104 (1), 47—57.

Tsokos, M., Schulz, F., 1999. Indoor postmortem animal interference by carnivores and rodents: report of two cases and review of the literature. Int. J. Leg. Med. 112 (2), 115—119.

SUPERORDER SELACHIMORPHA—SHARKS

Allaire, M.T., Manhein, M.H., Burgess, G.H., 2012. Shark-inflicted trauma: a case study of unidentified remains recovered from the Gulf of Mexico. J. Forensic Sci. 57 (6), 1675—1678.

Anderson, B., Manoukian, A., Holland, T., Grant, W., 2003. A death in paradise: human remains scavenged by a shark. In: Steadman, D. (Ed.), Hard Evidence: Case Studies in Forensic Anthropology, 1st ed Pearson Education, Inc, Upper Saddle River, pp. 186—196.

Byard, R.W., Gilbert, J.D., Brown, K., 2000. Pathologic features of fatal shark attacks. Am. J. Forensic Med. Pathol. 21 (3), 225—229.

Giertsen, J.C., Morild, I., 1989. Seafaring bodies. Am. J. Forensic Med. Pathol. 10 (1), 25—27.

Haglund, W.D., 1993. Disappearance of soft tissue and the disarticulation of human remains from aqueous environments. J. Forensic Sci. 38 (4), 806—815.

Ihama, Y., Ninomiya, K., Noguchi, M., Fuke, C., Miyazaki, T., 2009. Characteristic features of injuries due to shark attacks: a review of 12 cases. Leg. Med. 11 (5), 219—225.

Işcan, M.Y., McCabe, B.Q., 1995. Analysis of human remains recovered from a shark. Forensic Sci. Int. 72 (1), 15—23.

Klimley, A.P., 1994. The predatory behavior of the white shark. Am. Scient. 82 (2), 122—133.

Kringsholm, B., Jakobsen, J., Sejrsen, B., Gregersen, M., 2001. Unidentified bodies/skulls found in Danish waters in the period 1992—1996. Forensic Sci. Int. 123, 150—158.

Lowry, D., de Castro, A.L.F., Mara, K., Whitenack, L.B., Delius, B., Burgess, G.H., et al., 2009. Determining shark size from forensic analysis of bite damage. Mar. Biol. 156, 2483.

Rathbun, T., Rathbun, B., 1997. Human remains recovered from a shark's stomach in South Carolina. In: Haglund, W.D., Sorg, M.H. (Eds.), Forensic Taphonomy: The Postmortem Fate of Human Remains. CRC Press, Boca Raton, pp. 449—456.

Stock, M.K., Winburn, A.P., Burgess, G.H., 2017. Skeletal indicators of shark feeding on human remains: Evidence from Florida forensic anthropology cases. J. Forensic Sci. Available from: http://dx.doi.org/10.1111/1556-4029.13470.

Weng, K.C., Boustany, A.M., Pyle, P., Anderson, S.D., Brown, A., Block, B.A., 2007. Migration and habitat of white sharks (*Carcharodon carcharias*) in the Eastern Pacific Ocean. Mar. Biol. 152 (4), 877—894.

FAMILY SUIDAE—SUIDS—WILD BOAR AND DOMESTIC PIG

Berryman, H.E., 2001. Disarticulation pattern and tooth mark artifacts associated with pig scavenging of human remains: a case study. In: Haglund, W.D., Sorg, M.H. (Eds.), Advances in Forensic Taphonomy: Method, Theory, and Archaeological Perspectives. CRC Press, Boca Raton, pp. 487—495.

Cameron, S., 2011. On the Farm: Robert William Pickton and the Tragic Story of Vancouver's Missing Women. Vintage Canada, Toronto.

Corn, J.L., Warren, R.J., 1985. Seasonal food habits of the collard peccary in south Texas. J. Mammal. 66 (1), 155–159.

Domínguez-Solera, S.D., Domínguez-Rodrigo, M., 2009. A taphonomic study of bone modification and of tooth-mark patterns on long limb bone portions by suids. Int. J. Osteoarchaeol. 19 (3), 345–363.

Galdikas, B.M., 1978. Orangutan death and scavenging by pigs. Science 200 (4337), 68–70.

Greenfield, H.J., 1988. Bone consumption by pigs in a contemporary Serbian village: implications for the interpretation of prehistoric faunal assemblages. J. Field Archaeol. 15 (4), 473.

Mississippi State University Extension. 2017. Wild pig info. Available from: http://wildpiginfo. msstate.edu/history-wild-pigs.html (accessed 31.07.17).

Taylor, R.B., & Uvalde, T., 1999. Seasonal diets and food habits of feral swine. Proceedings of the First National Feral Swine Symposium, Ft. Worth, pp. 58–66.

FAMILY URSIDAE—URSIDS—BEARS

Atkinson, S., Ramsay, M., 1995. The effects of prolonged fasting of the body composition and reproductive success of female polar bears (Ursus maritimus). Funct. Ecol. 9 (4), 559–567.

Bright, L.N., 2011. Taphonomic signatures of animal scavenging in northern California: a forensic anthropological analysis. Unpublished Masters thesis. Department of Anthropology. California State University, Chico.

Carson, E.A., Stefan, V.H., Powell, J.F., 2000. Skeletal manifestations of bear scavenging. J. Forensic Sci. 45 (3), 515–526.

Cristescu, B., Stenhouse, G.B., Boyce, M.S., 2014. Grizzly bear ungulate consumption and the relevance of prey size to caching and meat sharing. Anim. Behav. 92, 133–142.

Elgmork, K., 1982. Caching behavior of brown bears (Ursus arctos). J. Mammal. 63 (4), 607–612.

Green, G.I., Mattson, D.J., Peek, J.M., 1997. Spring feeding on ungulate carcasses by grizzly bears in Yellowstone National Park. J. Wildl. Manage. 61 (4), 1040–1055.

Hilderbrand, G., Jenkins, S., Schwarts, C., Hanley, T., Robbins, C., 1999. Effect of seasonal differences in dietary meat intake on changes in body mass and composition in wild and captive brown bears. Can. J. Zool. 77 (10), 1623–1630.

Mowat, G., Heard, D.C., 2006. Major components of grizzly bear diet across North America. Can. J. Zool. 84 (3), 473–489.

Murad, T.A., Boddy, M.A., 1987. A case with bear facts. J. Forensic Sci. 32 (6), 1819–1826.

Nelson, R.A., Folk, G.E., Pfeiffer, E.W., Craighead, J.J., Jonkel, C.J., Steiger, D.L., 1983. Behavior, biochemistry, and hibernation in black, grizzly, and polar bears. Bears: Their Biology and Management 5, 284–290.

Saladié, P., Huguet, R., Díez, C., Rodríguez-Hidalgo, A., Carbonell, E., 2013. Taphonomic modifications produced by modern brown bears (Ursus arctos). Int. J. Osteoarchaeol. 23 (1), 13–33.

Thiemann, G.W., Iverson, S.J., Stirling, I., 2008. Polar bear diets and Arctic marine food webs: Insights from fatty acid analysis. Ecol. Monogr. 78 (4), 591–613.

Zager, P., Beecham, J., 2006. The role of American black bears and brown bears as predators on ungulates in North America. Ursus 17 (2), 95–108.

Pokines, J.T., Sussman, R., Gough, M., Ralston, C., McLeod, E., Brun, K., et al., 2017. Taphonomic analysis of Rodentia and Lagomorpha bone gnawing based upon incisor size. J. Forensic Sci. 62 (1), 50–66.

Compagno, L., Dando, M., Fowler, S., 2005. Sharks of the World. Princeton University Press, Princeton.

Ecological Influences on Scavenging Behavior

INTRODUCTION
ECOLOGICAL PRINCIPLES AND SCAVENGING BEHAVIOR
CLIMATE AND WEATHER
COMMUNITY COMPOSITION AND COMPETITION
ANTHROPOGENIC ACTIVITY
CONCLUSION: ECOLOGY, BEHAVIOR, AND TAPHONOMIC EFFECTS
REFERENCES

INTRODUCTION

Ecology is a branch of biology that focuses on associations between species and their environments, including their relationships with other living organisms and how they are influenced by environmental conditions. While scavenger morphology and physiology within a species have minor variation and remain within a predictable range, behavior can vary considerably between populations or at different times of the year. This intraspecies variation exists due to ecological variables including climate, habitat type, the composition of the local biological community (i.e., local species present), and even human activity. Such variables not only influence the probability of scavenging by a particular species within a community, but also produce differences in foraging behavior that can dramatically alter the evidence left at a death scene. For example, ecological context may affect soft tissue and underlying bones that are targeted by a scavenger, the rate at which these tissues are consumed, the direction that remains are scattered, or the probability of evidence removal or caching. Therefore a synthetic approach which considers known taphonomic signatures within the local ecological context is beneficial, given the flexible nature of animal

Forensic Taphonomy and Ecology of North American Scavengers. DOI: https://doi.org/10.1016/B978-0-12-813243-2.00006-3

behavior. This chapter introduces key ecological principles and discusses the potential effects of ecological variables on scavenger behavior.

ECOLOGICAL PRINCIPLES AND SCAVENGING BEHAVIOR

Some of the variation in scavenging frequency is inherent to a species, driven by adaptations that promote or preclude carrion detection and utilization. However, where and when animals search for food and the types of resources they ultimately utilize are not completely biologically predetermined, nor are they completely random. The day-to-day decisions animals make regarding foraging strategies, or the manner in which animals search for and obtain food sources in the wild, are continuously being shaped by their ecological context. Ecologists interested in animal decision-making have taken to modeling foraging scenarios, incorporating principles of economics to predict the types of habitats animals select, the types of resource patches they choose to exploit, and the amount of time they dedicate to a chosen patch (Stephens, 2008).

Pulliam (1974) identified four key variables which shape an animal's selection or rejection of particular food items: (1) encounter probability, (2) abundance of the food item, (3) handling time, and (4) energetic value of the food item. Encounter probability is the likelihood of locating a particular food item in a random search, and abundance describes the amount of the item available in the animal's environment; while handling time encompasses the time spent searching for, catching, consuming, and digesting the food item in question (Brown, 2009). Pulliam's model of diet choice (1974) predicts that animals will preferentially consume items that have higher ratios of energy gained to handling time. Further, they will consume less desirable items if the energy gained from them is greater than the net energy gained from a higher quality resource, *after* costs of searching and handling the high-quality item are considered (Brown, 2009).

Although simplistic, Pulliam's model (1974) is likely to explain the behavior of facultative scavengers, which are more likely to encounter carrion randomly rather than actively seek it. When preferred food items are abundant and easy to handle in the environment (e.g., a pulse of slow, inexperienced ungulate fawns following breeding season), facultative scavengers are less likely to utilize carrion that they encounter. However, if preferred foods are scarce (e.g., poor salmon runs due to downstream flooding that kills young fish), facultative scavengers have more to lose by rejecting carrion.

A second foraging dilemma that scavengers face is the decision about how extensively to utilize carrion once it has been located. Charnov (1976) proposed the Marginal Value Theorem (MVT) to explain how animals decide when to abandon feeding patches. The MVT states that a forager should leave a food patch when the amount of food remaining reaches a "giving-up density," indicating that the benefits that can be gained from the current patch are no longer greater than the average benefits available from any other patch in the environment (Brown, 2009). Consider a bear harvesting from a blackberry bush. When the bear arrives, the bush has yet to be exploited and is abundant with fruit. As the bear picks and eats the berries, the bush is gradually depleted. Berries near the surface of the bush will be exploited first, but as those are consumed the bear must search into deeper parts of the bush, under leaves and through branches, decreasing the bear's rate of harvest—or, as Pulliam (1974) might put it, increasing the handling time. Eventually the bush reaches the "giving-up density," and the bear leaves the bush to find another, where the energetic offerings are similar but the time and effort spent to access the fruit are lower.

Both Pulliam's (1974) model of diet choice and Charnov's (1976) MVT fall under the purview of optimal foraging theory, which suggests that the foraging behaviors of animals have evolved from strategies in the past that maximized energetic gains (McNamara and Houston, 1985). However, classical optimal foraging theory makes several false assumptions. First, optimal foraging theory assumes that animals have necessary knowledge of available resource patches to make informed foraging decisions; however, no animal begins with complete knowledge of a new environment (McNamara and Houston, 1985). In reality, learning about available resources plays an important role in maximizing foraging efficiency (McNamara and Houston, 1985). Learning is also critical for animals to adjust to environmental changes, whereas optimal foraging theory assumes an animal's environment is entirely static (McNamara and Houston, 1985). In addition, optimal foraging theory fails to account for the effects of competition and predation on foraging decisions, assuming that each organism in the biological community has unrestricted access to every available resource (Stephens, 2008).

A biological community is the collection of species that live within a defined region during a given time period, including all animals (vertebrate and invertebrate), as well as microorganisms, fungi, and plants. The composition of a biological community and the interactions between the organisms within it affect the likelihood that a given species will utilize a given resource. Within a biological community, food resources—including carrion—must be divided between the organisms that can consume them. Because these resources are finite, organisms from many different species are in competition with one another to obtain them, survive, and reproduce. Such competition may be intraspecific (i.e., between members of the same species) or interspecific (i.e., between organisms of different species).

Interspecific competition often occurs between species occupying the same ecological guild. Introduced by Root (1967), guilds are groups of species that fill a similar role in the community by exploiting similar resources (Simberloff and Dayan, 1991). Throughout this text, we have focused on the scavenging guild of North America. However, guilds may be structured around any resource, including food, shelter, or habitat. Consequently, species may be members of multiple guilds at once (Simberloff and Dayan, 1991). For example, in the American southeast, the American alligator (*Alligator mississippiensis*) belongs to the carnivorous reptile guild, the scavenging guild, and the mound-nesting guild.

As touched on previously in Chapter 3, competition can influence a population's evolutionary trajectory. Over many generations, morphological or physiological adaptations arise as existing beneficial traits are passed to offspring and become more common within a population, allowing its members to better compete with members of other species (as well as with conspecifics). However, differences in the behavior of populations can occur before physical distinctions evolve, because behavioral shifts are often the fastest way to respond to environmental changes that pose new challenges to population survival (Diogo, 2017; Wright et al., 2010). Without behavioral plasticity, or the ability to adapt behavior to new conditions, individuals would be vulnerable to even slight environmental change, and thus behavioral plasticity is adaptive in unstable environments (Wright et al., 2010). Behavior may be modified in response to countless interacting variables, but here we focus on how climate, community composition and competition, and anthropogenic activity can affect scavenging activity.

CLIMATE AND WEATHER

Climate plays a role in determining what scavenging species are present at a given location, as many species are well adapted to specific climes. The distributions of crocodilians, for example, are restricted to the southernmost regions of North America due to their status as ectotherms (i.e., they are cold-blooded, meaning their body temperature is regulated by the external environment) while the range of the polar bear (*Ursus maritimus*), adapted to extreme cold, is restricted to the circumpolar Arctic. Other species migrate seasonally to reduce exposure to climatic variation. For example, the range of turkey vultures extends northward and westward in the summer; across the continent, populations of turkey vultures can be found as far north as Canada in the summer or as far south as Paraguay in the winter (Kiff, 2000).

Other scavenging species occupy the same territories year-round, but alter their use of land and prey resources with changing seasons. In some parts of Idaho, coyotes (*Canis latrans*), bobcats (*Lynx rufus*), and pumas (*Puma concolor*) coexist as both predators and scavengers. During the favorable summer months, the three species reduce competition for their shared food resources by foraging at varying elevations, a strategy known as resource partitioning, while extreme winter weather forces all three species to migrate to lower, warmer elevations (Koehler and Hornocker, 1991).

Local weather can also influence the probability of a species opting to scavenge. Higher temperatures accelerate decomposition by catalyzing cellular autolysis (i.e., the breakdown of cell membranes) and microbial activity, the latter of which produces olfactory cues that serve as the initial attraction for both vertebrate and invertebrate scavengers (Beasley et al., 2015; Campobasso et al., 2001; Tomberlin et al., 2012). Most facultative scavengers have a preference for fresh carrion and are more likely to scavenge when temperatures are low because this slows microbial reproduction and invertebrate colonization, expanding the vertebrate scavenging window (DeVault et al., 2004; Selva et al., 2005). Alternatively, warmer temperatures will largely restrict scavenging activity to that of species which are more tolerant of invertebrates and the byproducts of advanced decomposition, such as vultures (Cathartidae spp.) (Young et al., 2014).

Environmental temperature is also an important variable that affects the timing and duration of scavenging by vertebrates. High daily temperatures will decrease the time between death and discovery of carrion by accelerating the decomposition rate, expediting the release of the attractive odors of microbial metabolism (Tomberlin et al., 2012). This also leads to rapid insect involvement with the carcass, blow flies (Calliphoridae spp.) in particular. Therefore high temperatures narrow the vertebrate scavenging window for many species. For example, raccoons (*Procyon lotor*) discover remains quickly after death but cease scavenging sooner in the summer when temperatures are warmer (Hannigan, 2015; Jeong et al., 2016). If carrion is not discovered during the vertebrate scavenging window, high temperatures may prevent many vertebrate species from scavenging at all due to their aversion to the byproducts of advanced decomposition.

Some species will attempt to mitigate the effects of higher temperatures, employing strategies to reduce spoilage and avoid competition with microorganisms and invertebrates. An experimental study conducted in Arizona by Bischoff-Mattson and Mattson (2009) replicated the caching behavior of pumas (*Puma concolor*) by depositing carcasses of ungulate species in a shady location under a thin layer of soil and organic debris. The temperature, rate of decomposition, and odor emanation of experimental carcasses were significantly reduced compared to control counterparts that were deposited uncovered in the sun. Other evidence of caching behaviors that reduce decomposition rates has been reported. For example, in Norway, brown bears (*Ursus arctos*) cache carcasses under large mounds of vegetation that frequently include a species of moss with demonstrated antimicrobial and antifungal properties (Elgmork, 1982).

In addition to temperature, precipitation may affect scavenging activity in some species. Significant rainfall tends to inhibit scavenging by both vultures and terrestrial scavengers (Beck et al., 2015; Jones, 2011; Reeves, 2009; Selva et al., 2005). Further north, deep snow reduces the proportion of carrion in the winter diets of large carnivores because hunting success rates are improved. For example, gray wolves (*Canis lupus*) in winter have been shown to supplement their diets with scavenging in periods of little snowfall but rely on hunting when snow cover is high, because deeper snow hinders the escape of ungulate prey (Huggard, 1993). Hunting is the foraging strategy of choice for wolves when the success rate is high because (1) energetic costs are offset and (2) they prefer fresh tissue to carrion. In contrast, corvids (Corvidae spp.) have been shown to increase utilization of carcasses in winter, perhaps due in part to the reduced competition from larger avian and terrestrial species (Selva et al., 2005).

Anthropogenic Climate Change

It is outside the scope of this text to discuss the extent of influence that anthropogenic activity has on global climate. A brief review of literature from the early 21st century indicates that most Americans are aware of climate change, although levels of public concern remain consistently low and acceptance of a role for human activity in shaping these changes is among the lowest in the world (Brulle et al., 2012; Leiserowitz, 2005; Semenza et al., 2008; Weber and Stern, 2011). However, the overwhelming consensus of the scientific community has long supported an anthropogenic component to climate change (Anderegg et al., 2010; Oreskes, 2004).

Global temperatures have fluctuated over the last several decades, simultaneously affecting patterns of humidity, wind, and precipitation. In the northern hemisphere, temperatures have exhibited a general warming trend (Mann et al., 1999). Given the influential role of climate in shaping community composition discussed previously in this chapter, climate change poses a threat to the stability and functioning of existing ecological communities (Walther et al., 2002). Individual species, especially those with restricted ranges, are at a heightened risk for extinction (Maclean and Wilson, 2011). Scavengers are no exception and, as we will soon learn, changes in the health or behavior of one species have cascading effects on the health and behavior of many other species.

Climate change has a multitude of ecological effects, many which seem relatively benign but nevertheless threaten the structure of scavenging communities. With changing weather patterns and corresponding shifts in vegetation and/or prey, species that are adapted to particular climates may be forced to shift their ranges with varying levels of success; the tragic plight of the polar bear is perhaps the most familiar example (Hunter et al., 2010; Parmesan, 2006). In addition, microbial and insect activity is promoted by the warming global climate, improving conditions for invertebrate pests and transmission of bacterial diseases (Parmesan, 2006); these changes have troubling consequences for all organisms, including humans. For facultative scavengers, the promotion of microbial and invertebrate activity would have the additional detrimental effect of narrowing the vertebrate scavenging window (Beasley et al., 2015; DeVault et al., 2003).

Mid-sized predators that previously relied on large carrion produced by winter starvation and disease, like the coyote, and mesopredators like the raccoon, may struggle to meet energetic demands as winters become more mild, reducing the efficiency of acquiring large prey (Wilmers and Getz, 2005). Models of climate-change scenarios have suggested that the presence of apex predators may somewhat buffer these losses via their abandonment of large kills, which produces a consistent supply of carrion (Wilmers and Getz, 2005; Wilmers and Post, 2006). However, the effects of climate change on obligate scavengers have yet to be modeled, although it is not unreasonable to predict that current vulture populations may benefit from a warming climate. Vultures have previously been excluded from temperate regions because their primary mechanism of locomotion—a critical contributor to their survival as obligate scavengers—relies on uplift generated by thermal currents, which are stronger in warmer climate zones (Dodge et al., 2014). However, in the past several decades temperatures in North America have increased and New World vulture populations have expanded their ranges northward (Kiff, 2000). Coupled with the carrion pulses produced by multiple predicted species extinctions, vultures may be among the few species to emerge victorious from climate change scenarios.

COMMUNITY COMPOSITION AND COMPETITION

The composition of a biological community, or the species represented, can impact the behavior of scavengers in numerous ways. The nature of the relationships between each represented species are complex and include predator–prey, competitive, and cooperative dynamics (Amarasekare, 2009; Bronstein, 2009; Denno and Lewis, 2009). Consequently, changes in the abundance of any one species can have a cascading effect. For example, suppose a northern forest sees a particularly cold winter, which kills the buds of a native species of flowering plant that has characterized the region for decades. Come spring, the flowers of most of these plants fail to bloom. The flower's nectar is a critical food resource for several species of insect and provides a dietary supplement to others, so these populations suffer. Many bird species which feed on these insects also feel the effect. By summer, only a portion of the plants have been pollinated as they are few and far between, and consequently very few of the plants produce seeds.

Several rodent species in the community use these seeds in their diet and incorporate them into their winter caches. While the produced seeds are collected by the rodents, those that are consumed or successfully cached cannot produce another plant in the spring. Despite attempting to supplement caches with other seed species, the rodent population suffers when winter comes, and subsequently the rodents' predators—including predatory birds and terrestrial predators, such as coyotes—also struggle to meet their nutritional needs with the decline of both the bird and rodent populations. With the decline of the coyote population, the ungulates thrive the following spring, but overgraze vegetation that is an important staple in the diet of local rabbit populations. Therefore what started as a relatively minor disruption to a single species can have negative or positive effects on the entire food chain that can be felt for years afterward.

A well-known example that serves to further reinforce the point is that of the effect gray wolf reintroduction to Yellowstone National Park in the United States has had on the ecosystem. Absent for 70 years after being eradicated, they were reintroduced in 1995 as part of conservation efforts (Ripple and Beschta, 2012). Since then, their presence has led to a trophic cascade which began with deer changing their behavior to avoid the wolves, leading to regeneration of trees and other vegetation that the deer had been browsing on in those areas (Ripple et al., 2001), and eventually causing a decrease to the elk population and increases to the beaver and bison populations, both positive developments for the ecosystem (Ripple and Beschta, 2012). Further, in the absence of the wolf, increased deer browsing on vegetation along streams in the park actually contributed to geological changes in their width and hydraulic capacity (i.e., decreased efficiency of the floodplain in modulating flood flows), with detrimental consequences for local plants and other animals (Beschta and Ripple, 2006). In addition, a study by Wilmers and Getz (2005) demonstrated that wolves mitigate the effect of climate change (i.e., shorter winters) on animals that scavenge to survive the winter by providing ample carrion. Their model demonstrates that earlier snowmelt in ecosystems without wolves have a reduction in late winter carrion which has consequences for scavenging species (Wilmers and Getz, 2005).

Hunting Success

Declines in various animal populations produce carrion, which appear as resource pulses (Ostfeld and Keesing, 2000). The general health of populations determines how much carrion is deposited and can also impact how frequently particular species scavenge. For apex predators (i.e., wolves and bears), the amount of carrion included in the diet is heavily dependent on hunting success. Hunting is an energetically expensive activity, so when success rates are low, scavengers may use carrion to supplement their diet. Conversely, poor health of prey populations due to food shortages or disease outbreaks increases the hunting success of their predators, leading to decreased utilization of carrion by the predatory species (Metz et al., 2012; Stahler et al., 2006).

Therefore, increased reliance on hunting strategies by predators has the potential to increase the amount of carrion available in the general environment, especially in conjunction with poor prey population health. It is difficult to predict how the biological community may respond to increases in carrion availability and changes in competitor scavenging frequency. However, Olson et al. (2012) suggest that other vertebrates will increase scavenging efforts when a dominant competitor is removed from the community. Their study examined the systematic removal of a raccoon population from a site in Indiana and found that other vertebrate scavengers, such as opossums (*Didelphis virginiana*), became more active in the absence of their primary competitors (Olson et al., 2012). Surprisingly, this study found no increase in invertebrate scavenging (Olson et al., 2012). It is possible that scavenging guild members may respond similarly in other communities where the dominant species persists but significantly reduces their utilization of carrion.

The Influence of Reproduction on Predators and Prey
Hunting success is also influenced by the reproductive cycles of prey, as pregnant females and juveniles are more vulnerable to predation (Metz et al., 2012; Stahler et al., 2006). However, reproductive habits of predatory species may also affect how carrion is utilized. In some taxa, including bears (*Ursus* spp.), daily movement increases as animals actively search for mates, increasing the likelihood of carcass discovery and scavenging (Krofel et al., 2012).

Pregnancy and lactation are energetically expensive for females and may require caloric supplementation. As discussed in Chapter 5, bears experience dietary shifts during hyperphagia (i.e., hibernation preparation), which precedes gestation, birth, and lactation (Bright, 2011). In Yellowstone National Park, peak biomass acquisition—including hunted and scavenged carcasses—by gray wolves occurred in late winter and early spring, coinciding with the wolf's breeding and whelping season (Metz et al., 2012). While Metz et al. (2012) did not demonstrate seasonal differences in scavenging frequency, the wolf's breeding season begins in late winter when ungulate populations are weakened and snowfall tends to be high. This promotes hunting success and thereby alleviates the pressures of heightened energy needs (Huggard, 1993; Metz et al., 2012; Wilmers and Getz, 2005).

Competition Between Species
In addition to predator–prey dynamics, competitive relationships between scavenging species also impact behavior. Given the highly competitive nature of carrion acquisition, scavenging guilds have a nested structure (Sebastián-González et al., 2016; Selva and Fortuna, 2007). This means that a predictable "core" set of species scavenge most carcasses, including those scavenged by only a few species, while animals that rarely scavenge are only found at carcasses scavenged by a large number of species. This is in contrast to scavenging of most carcasses by a random assortment of scavenging species (Selva and Fortuna, 2007). For example, most carrion deposited in the American southwest will be utilized by vultures at some point, with some monopolized by the species; in contrast, mesopredators (e.g., medium-sized predators such as foxes, raccoons, and opossums) may only be

documented at higher competition carcasses that are also exploited by several other species.

Nestedness implies that the community is subject to a high degree of interspecific competition, with species highly adapted to carrion acquisition able to effectively prevent other species from accessing a carcass. However, variation in nestedness and the composition of the core scavenging species is associated with carcass size, with small carcasses able to be controlled by mesopredators (Moleón et al., 2014). For larger carcasses, the community's degree of nestedness is also strongly influenced by the presence of obligate or other highly specialized scavengers (Moleón et al., 2014; Sebastián-González et al., 2013; Selva and Fortuna, 2007).

In scavenging guilds with higher richness (i.e., greater number of scavenging species), the time between death and carcass consumption is reduced due to the increased likelihood of rapid detection (Sebastián-González et al., 2016). The presence of specialized scavengers in a region further reduces this interval (Sebastián-González et al., 2013). Scavenging community richness and the presence of specialized scavengers may have predictable effects on decomposition rate, suggesting that it may be possible to calibrate postmortem interval estimates made on scavenged remains based on evidence of active scavenging taxa and the characteristics of the biological community. This hypothesis, however, has yet to be explored.

Increased richness of the scavenging guild increases interspecific competition for carrion, as more species are competing for the same resources. Dominance hierarchies limit the number of species foraging from a given carcass, with more dominant species excluding the presence of subordinate ones. This contributes to the nested nature of scavenging guilds. Obligate scavengers and apex predators are generally well equipped to exclude other species from carcasses, although the methods they employ vary. As discussed in Chapter 5, vultures are highly voracious scavengers that can easily strip a carcass of tissue in several hours of active feeding, effectively precluding scavenging by competitors (Reeves, 2009; Spradley et al., 2012). The practice of foraging in large groups, referred to as "wakes" of vultures, further promotes exclusion of competitors, including apex predators (Allen et al., 2015; Elbroch and Wittmer, 2013).

In contrast, apex predators rely on their superior size and strength to reduce competition. Bears tend to hold the dominant position in the scavenging guild, limiting carrion access of their competitors, including large felids (Allen et al., 2015; Elgmork, 1982; Krofel et al., 2012). Felids, in turn, are typically able to preclude carrion acquisition by bird species and mesopredators (Allen et al., 2015). It should be noted, however, that dominance does not necessarily guarantee protection against carcass usurpation. One interesting report documents a Western spotted skunk (*Spilogale gracilis*) successfully usurping and defending a deer carcass killed by a puma (Allen et al., 2013).

While research on competitive interactions between canids and other large carnivores is scarce, canids are likely at an advantage given their propensity of hunting in packs, a form of social foraging that promotes rapid carrion consumption as well as defense (Atwood and Gese, 2008; Paquet, 1992). For canids, pack size influences the success of carrion acquisition and defense. For example, Atwood and Gese (2008) demonstrated that coyotes foraging in larger packs were more successful in usurping the kills of bigger, dominant gray wolves. The importance of social foraging in canids is also demonstrated by positive correlations between coyote pack size and the amount of meat in the population's diet, as well as increased pack size in winter months, when scavenging activity increases to compensate for winter food shortages (Bowen, 1981; Gese et al., 1988).

Vertebrate scavengers respond to competition in many ways. Generalist species may utilize carrion indirectly by preying on associated invertebrates. Raccoons, for example, will prey on maggot larvae colonizing a carcass when tissue is too decomposed for direct consumption (Jeong et al., 2016). Corvids and passerine birds (i.e., perching birds), the latter of which do not consume tissue, will also feed on invertebrates associated with carrion (Moreno-Opo and Margalida, 2013).

Many species will utilize competitors to increase the efficiency of their carrion consumption, engaging in kleptoparasitism or food theft. For example, American black vultures (*Coragyps atratus*) will usurp carcasses located by turkey vultures (*Cathartes aura*), reducing their own search efforts by taking advantage of their competitor's better olfaction (Wallace and Temple, 1987). Additionally, it is well known that many avian and mammalian scavengers will purposefully associate with higher-level predators to utilize abandoned carcasses and their neglected tissue scraps (Arjo and Pletscher, 2000; Heinrich and Pepper, 1998; Kane et al., 2014; King et al., 2016; Morton and Lord, 2006; Stahler et al., 2002; Switalski, 2003). This is a particularly lucrative strategy for species such as raven (*Corvus corax*) or opossum (*Didelphis virginiana*) that are incapable of breaking through the tough hides of larger carcasses to access internal tissues and are dependent on natural orifices or openings created by other species (Morton and Lord, 2006; Stahler et al., 2002).

Alternatively, species may shift their foraging habits strategically to avoid competition, partitioning resources to promote coexistence. Competition by turkey vultures and black vultures is reduced by adapting carrion selection, with turkey vultures typically specializing on small carrion and black vultures preferentially feeding on large carcasses, although both species will utilize carrion of any size if discovered (Buckley, 1996). Other than adjusting carcass selection, competitors may separate themselves by dividing foraging territories, thus avoiding the threat of kleptoparasitism. Bears, for example, can avoid competition by foraging at different elevations than competing species (Allen et al., 2015; Bankaitis, 2012; Green et al., 1997; Koehler and Hornocker, 1991). Wolf packs, which are highly territorial, avoid foraging in areas of territory overlap to reduce intraspecific conflict, creating "buffer zones" that are utilized as foraging sites by other competitors, including foxes (*Vulpes* and *Urocyon* spp.) and coyotes (Switalski, 2003). In areas inhabited by vultures, some large carnivores may avoid foraging in open habitats, preferring instead the cover of forested terrain that allows carcass concealment (Elbroch and Wittmer, 2013).

Caching is another strategy employed by many species to reduce the risk of carrion loss to competitors. Most New World vultures do not cache, although California condors (*Gymnogyps califorianus*) are known to hoard small skeletal elements in their nests as a calcium source (Collins et al., 2000). Corvids, in contrast, are notorious hoarders. Small bones and bone fragments have been recovered in black-billed magpie (*Pica hudsonia*) nests hundreds of meters from an experimental decomposition site (Komar and Beattie, 1998). In addition, common ravens (*Corvus corax*) are scatter-hoarders, caching small amounts of food in numerous places, often flown over significant distances to prevent cache theft by conspecifics (Heinrich and Pepper, 1998).

The caches of terrestrial carnivores are usually constructed in areas of dense vegetation, decreasing the likelihood of the carcass being detected by a competitor. Canids cache via active burial, digging holes before covering remains with loose soil, plant litter, or snow. Red foxes (*Vulpes vulpes*) are scatter-hoarders and tend to cache numerous small items, while larger canids may cache entire carcasses (Macdonald, 1976). Domestic dogs (*Canis familiaris*) are well known for their propensity to dig holes and bury food; and gray wolves have been observed burying an intact deer carcass under snow, returning to consume it within a week (Nelson, 2011). Rather than bury remains, bears and felids will cache by scraping available debris into a mound above the carcass. Bear caches can be distinguished from felid caches by their fairly circular shape and greater volume of covering material (Elgmork, 1982). Further, felid caches often incorporate the fur or hair of the carcass into the cache; this inclusion has been demonstrated in a case of bobcat caching of human remains (Rippley et al., 2012; Vander Wall, 1990).

ANTHROPOGENIC ACTIVITY

Anthropogenic activity can have major impacts on the behavior of scavengers. Generally, humans serve as a deterrent to scavengers. Carnivore scavenging is far more common in areas of low human population density (Haglund et al., 1989). In addition, scattering patterns often reveal that animals have deliberately dragged carcasses away from sites of human activity such as farms, roadways, or residential communities (Kjorlien et al., 2009; Ricketts, 2013). For example, grizzly bears (*Ursus arctos*) in Yellowstone National Park are less likely to utilize carcasses located near major park roads or recreational developments (Green et al., 1997).

Agricultural or residential development results in the fragmentation of habitats for many species and reduces the likelihood of carrion exploitation due to the repellent effect of human activity. Admittedly however, such patterns of carrion utilization may be mediated by the type of carcass available (Olson et al., 2016). It is important to note that not all scavenging species are averse to humans; indeed many thrive in anthropogenic environments. This includes domesticated dogs and cats, several species of rodent, and mesopredator pests like raccoons and opossums. In agricultural landscapes, mesopredators dominate the scavenging guild, although turkey vultures have been reported to take advantage of pests that have been trapped, killed, and buried by farmers seeking to protect their crops (DeVault et al., 2011; Smith et al., 2002).

The introduction of invasive scavenging species, or reintroduction of native species as part of conservation efforts, may affect scavenger behavior by altering community composition. As discussed earlier, gray wolves were reintroduced to Yellowstone National Park in the mid-1990s after having been exterminated from the region in the early 20th century (Switalski, 2003). The reintroduction provides a continuous source of carrion for animals lower on the food chain and in fact increases coyote scavenging frequency; but has also caused a significant decline in the coyote population due to fatal conflicts with wolves defending their kills (Switalski, 2003). In addition, coyote territories have become disproportionately located in the "buffer zones" where wolf territories overlap, as wolf packs avoid these areas to minimize intraspecific conflict (Switalski, 2003). Similar behavioral changes were seen in coyotes responding to the reintroduction of wolves in northwestern Montana, in addition to creating a temporal boundary by shifting toward nocturnal activity patterns (Arjo and Pletscher, 2000).

In addition, hunting by humans has a variable influence on carrion availability in the environment, with the effect on scavenger guilds dependent on the cultural characteristics of the hunt itself. Hunting of small game and mid-size deer is relatively popular in North America, and hunters will often remove and use the entire carcass. When this occurs, biomass is removed from the community to the detriment of predators and scavengers alike (Wilson and Wolkovich, 2011). If carcasses are gutted or only body parts are removed, as is often the practice of big game or "trophy" hunters, the hunt then subsidizes the scavenging community through the production of carrion (Mateo-Tomás et al., 2015; Wilson and Wolkovich, 2011). Consumption of hunted carrion is dominated by social birds (e.g., corvids) and generalist mammals (e.g., suids or mesopredators) in regions where obligate scavengers are absent and scavenging by apex predators is low. Therefore hunting has the potential to alter community structure by promoting population growth of those species taking advantage of the resource (Mateo-Tomás et al., 2015).

Research has also demonstrated that hunter-provided carrion in the industrialized world has a different spacio-temporal distribution than carrion provided by predator kills, because hunting is often restricted to particular seasons and locations. Therefore this produces carrion deposits that are spatially and temporally clumped (Wilmers et al., 2003). In a comparison of hunter-produced carrion and wolf-killed carcasses in Yellowstone National Park, Wilmers et al. (2003) noted that only corvids and predatory eagles (Accipitridae spp.) were documented at hunter-produced carrion deposits, while wolf-killed carcasses attracted a greater diversity of species and were largely dominated by coyotes. Further, obligate scavengers were notably absent from wolf kills as opposed to hunted carrion, although turkey vultures were occasionally observed at wolf kills in summer months.

Although some of the variation can be attributed to seasonal differences in sampling periods, Wilmers et al. (2003) suggest that these differences may also reflect the adaptive value of the high foraging radii of avian species. This allows them to recruit large groups of conspecifics to compete with terrestrial species at the highly clumped hunter-provided carrion (Wilmers et al., 2003). Coyote avoidance of human activity was also hypothesized to contribute to their complete exclusion from the hunter-provided carrion (Wilmers et al., 2003). It should be noted that with the exception of mass fatality events, human remains are likely to be dispersed in a manner similar to wolf kills, and therefore will likely attract a comparable diversity of species.

CONCLUSION: ECOLOGY, BEHAVIOR, AND TAPHONOMIC EFFECTS

The goal of this chapter was to introduce basic principles of ecology and review key literature on how a few ecological variables—climate, community composition, and anthropogenic activity—can influence patterns of carcass utilization and movement, resulting in deviations from species-typical taphonomic effects. The variables discussed here are not the only ecological variables that may come into play, nor are their effects mutually exclusive. However, the point is that ecological variables interact to influence animal behavior. The complexity of ecology on scavenging behavior therefore necessitates a case-by-case approach to interpreting a scavenged death scene, which is outlined in the following chapter.

REFERENCES

Allen, M.L., Elbroch, L.M., Wilmers, C.C., Wittmer, H.U., 2015. The comparative effects of large carnivores on the acquisition of carrion by scavengers. Am. Nat. 185 (6), 822–833.

Allen, M.L., Elbroch, L.M., Wittmer, H.U., 2013. Encounter competition between a cougar, *Puma concolor*, and a western spotted skunk, *Spilogale gracilis*. Can. Field Nat. 127 (1), 64–66.

Amarasekare, P., 2009. Competition and coexistence in animal communities. In: Levin, S.A. (Ed.), The Princeton Guide to Ecology. Princeton University Press, Princeton, pp. 196–201.

Anderegg, W.R., Prall, J.W., Harold, J., Schneider, S.H., 2010. Expert credibility in climate change. Proc. Natl Acad. Sci. 107 (27), 12107–12109.

Arjo, W.M., Pletscher, D.H., 2000. Behavioral responses of coyotes to wolf recolonization in northwestern Montana. Can. J. Zool. 77 (12), 1919–1927.

Atwood, T.C., Gese, E.M., 2008. Coyotes and recolonizing wolves: social rank mediates risk-conditional behaviour at ungulate carcasses. Anim. Behav. 75 (3), 753–762.

Bankaitis, J., 2012. Examination of scavenging associated with wolves. Unpublished Masters thesis. Department of Anthropology. The University of Montana, Missoula.

Beasley, J.C., Olson, Z.H., DeVault, T.L., 2015. Ecological role of vertebrate scavengers. In: Benbow, M.E., Tomberlin, J.K., Tarone, A.M. (Eds.), Carrion Ecology, Evolution and Their Applications. CRC Press, Boca Raton, pp. 107–127.

Beck, J., Ostericher, I., Sollish, G., De Leon, J., 2015. Animal scavenging and scattering and the implications for documenting the deaths of undocumented border crossers in the Sonoran Desert. J. Forensic Sci. 60 (S1), S11–S20.

Beschta, R., Ripple, W., 2006. River channel dynamics following extirpation of gray wolves from Yellowstone National Park, USA. Earth Surf. Proc. Land. 31 (12), 1525–1539.

Bischoff-Mattson, Z., Mattson, D., 2009. Effects of simulated mountain lion caching on decomposition of ungulate carcasses. West. North Am. Nat. 69 (3), 343–350.

Bowen, W.D., 1981. Variation in coyote social organization: the influence of prey size. Can. J. Zool. 59 (4), 639–652.

Bright, L.N., 2011. Taphonomic signatures of animal scavenging in northern California: a forensic anthropological analysis. Unpublished Masters thesis. Department of Anthropology. California State University, Chico.

Bronstein, J., 2009. Mutualism and symbiosis. In: Levin, S.A. (Ed.), The Princeton Guide to Ecology. Princeton University Press, Princeton, pp. 233–238.

Brown, J.S., 2009. Foraging behavior. In: Levin, S. (Ed.), The Princeton Guide to Ecology. Princeton University Press, Princeton, pp. 51–58.

Brulle, R.J., Carmichael, J., Jenkins, J.C., 2012. Shifting public opinion on climate change: an empirical assessment of factors influencing concern over climate change in the US, 2002–2010. Clim. Change 114 (2), 169–188.

Buckley, N.J., 1996. Food finding and the influence of information, local enhancement, and communal roosting on foraging success of North American vultures. Auk 113 (2), 473–488.

Campobasso, C.P., Di Vella, G., Introna, F., 2001. Factors affecting decomposition and Diptera colonization. Forensic Sci. Int. 120, 18–27.

Charnov, E.L., 1976. Optimal foraging, the marginal value theorem. Theoret. Popul. Biol. 9 (2), 129–136.

Collins, P.W., Noel, F.R.S., Emslie, S.D., 2000. Faunal remains in California Condor nest caves. Condor 102 (1), 222–227.

Denno, R., Lewis, D., 2009. Predator-prey interactions. In: Levin, S.A. (Ed.), The Princeton Guide to Ecology. Princeton University Press, Princeton, pp. 202–212.

DeVault, T.L., Brisbin Jr, I.L., Rhodes Jr, O.E., 2004. Factors influencing the acquisition of rodent carrion by vertebrate scavengers and decomposers. Can. J. Zool. 82 (3), 502–509.

DeVault, T.L., Olson, Z.H., Beasley, J.C., Rhodes Jr, O.E., 2011. Mesopredators dominate competition for carrion in an agricultural landscape. Basic Appl. Ecol. 12 (3), 268–274.

DeVault, T.L., Rhodes, O.E., Shivik, J.A., 2003. Scavenging by vertebrates: behavioral, ecological, and evolutionary perspectives on an important energy transfer pathway in terrestrial ecosystems. Oikos 102 (2), 225–234.

Diogo, R., 2017. Baldwin's organic selection and the increasing awareness of the evolutionary importance of behavioral shifts. In: Diogo, R. (Ed.), Evolution Driven by Organismal Behavior: A Unifying View of Life, Function, Form, Mismatches and Trends. Springer International Publishing, Cham, pp. 25–33.

Dodge, S., Bohrer, G., Bildstein, K., Davidson, S.C., Weinzierl, R., Bechard, M.J., et al., 2014. Environmental drivers of variability in the movement ecology of Turkey Vultures (*Cathartes aura*) in North and South America. Philos. Trans. R. Soc. B Biol. Sci. 369 (1643), 17.

Elbroch, L.M., Wittmer, H.U., 2013. Nuisance ecology: do scavenging condors exact foraging costs on pumas in Patagonia? PLoS One 8 (1), 1–8.

Elgmork, K., 1982. Caching behavior of brown bears (*Ursus arctos*). J. Mammal. 63 (4), 607–612.

Gese, E.M., Rongstad, O.J., Mytton, W.R., 1988. Relationship between coyote group size and diet in southeastern Colorado. J. Wildl. Manage. 52 (4), 647–653.

Green, G.I., Mattson, D.J., Peek, J.M., 1997. Spring feeding on ungulate carcasses by grizzly bears in Yellowstone National Park. J. Wildl. Manage. 61 (4), 1040–1055.

Haglund, W.D., Reay, D.T., Swindler, D.R., 1989. Canid scavenging/disarticulation sequence of human remains in the Pacific Northwest. J. Forensic Sci. 34 (3), 587–606.

Hannigan, A., 2015. A descriptive study of forensic implications of raccoon scavenging in Maine. Unpublished Honors thesis. Department of Anthropology. University of Maine, Orono.

Heinrich, B., Pepper, J.W., 1998. Influence of competitors on caching behaviour in the common raven, *Corvus corax*. Anim. Behav. 56 (5), 1083–1090.

Huggard, D.J., 1993. Effect of snow depth on predation and scavenging by gray wolves. J. Wildl. Manage. 57 (2), 382–388.

Hunter, C.M., Caswell, H., Runge, M.C., Regehr, E.V., Amstrup, S.C., Stirling, I., 2010. Climate change threatens polar bear populations: a stochastic demographic analysis. Ecology 91 (10), 2883–2897.

Jeong, Y., Jantz, L.M., Smith, J., 2016. Investigation into seasonal scavenging patterns of raccoons on human decomposition. J. Forensic Sci. 61 (2), 467–471.

Jones, A.L., 2011. Animal scavengers as agents of decomposition: the postmortem succession of Louisiana wildlife. Unpublished Masters thesis. Department of Geography and Anthropology. Louisiana State University, Baton Rouge.

Kane, A., Jackson, A.L., Ogada, D.L., Monadjem, A., McNally, L., 2014. Vultures acquire information on carcass location from scavenging eagles. Proc. R. Soc. B Biol. Sci. Available from: http://dx.doi.org/10.1098/rspb.2014.1072.

Kiff, L.F. 2000. The current status of North American vultures. In: Chancellor, R.D., & Meyburg, B. (Eds.), Raptors at Risk: Proceedings of the 5th World Conference on Birds of Prey & Owls. Hancock House, Towcester, pp. 175–189.

King, K.A., Lord, W.D., Ketchum, H.R., O'Brien, R.C., 2016. Postmortem scavenging by the Virginia opossum (*Didelphis virginiana*): impact on taphonomic assemblages and progression. Forensic Sci. Int. 266 (576), e1-576.e6.

Kjorlien, Y.P., Beattie, O.B., Peterson, A.E., 2009. Scavenging activity can produce predictable patterns in surface skeletal remains scattering: observations and comments from two experiments. Forensic Sci. Int. 188, 103–106.

Koehler, G.M., Hornocker, M.G., 1991. Seasonal resource use among mountain lions, bobcats, and coyotes. J. Mammal. 72 (2), 391–396.

Komar, D., Beattie, O., 1998. Identifying bird scavenging in fleshed and dry remains. J. Can. Soc. Forensic Sci. 31 (3), 177–188.

Krofel, M., Kos, I., Jerina, K., 2012. The noble cats and the big bad scavengers: effects of dominant scavengers on solitary predators. Behav. Ecol. Sociobiol. 66 (9), 1297–1304.

Leiserowitz, A.A., 2005. American risk perceptions: is climate change dangerous? Risk Anal. 25 (6), 1433–1442.

Macdonald, D.W., 1976. Food caching by red foxes and some other carnivores. Z. Tierpsychol. 42 (2), 170–185.

Maclean, I.M.D., Wilson, R.J., 2011. Recent ecological responses to climate change support predictions of high extinction risk. Proc. Natl Acad. Sci. U. S. A. 108 (30), 12337–12342.

Mann, M.E., Bradley, R.S., Hughes, M.K., 1999. Northern hemisphere temperatures during the past millennium: inferences, uncertainties, and limitations. Geophys. Res. Lett. 26 (6), 759–762.

Mateo-Tomás, P., Olea, P.P., Moleón, M., Vicente, J., Botella, F., Selva, N., et al., 2015. From regional to global patterns in vertebrate scavenger communities subsidized by big game hunting. Divers. Distrib. 21 (8), 913–924.

McNamara, J.M., Houston, A.I., 1985. Optimal foraging and learning. J. Theoret. Biol. 117 (2), 231–249.

Metz, M.C., Smith, D.W., Vucetich, J.A., Stahler, D.R., Peterson, R.O., 2012. Seasonal patterns of predation for gray wolves in the multi-prey system of Yellowstone National Park. J., Anim. Ecol. 81 (3), 553–563.

Moleón, M., Sánchez-Zapata, J.A., Selva, N., Donázar, J.A., Owen-Smith, N., 2014. Inter-specific interactions linking predation and scavenging in terrestrial vertebrate assemblages. Biol. Rev. 89 (4), 1042–1054.

Morton, R.J., Lord, W.D., 2006. Taphonomy of child-sized remains: a study of scattering and scavenging in Virginia, USA. J. Forensic Sci. 51 (3), 475–479.

Moreno-Opo, R., Margalida, A., 2013. Carcasses provide resources not exclusively to scavengers: patterns of carrion exploitation by passerine birds. Ecosphere 4 (8), 1–15.

Nelson, M.E., 2011. Killing and caching of an adult white-tailed deer, Odocoileus virginianus, by a single gray wolf, Canis lupus. Can. Field Nat. 125 (2), 162–164.

Olson, Z., Beasley, J., DeVault, T.L., Rhodes, O., 2012. Scavenger community response to the removal of a dominant scavenger. Oikos 121 (1), 77–84.

Olson, Z.H., Beasley, J.C., Rhodes Jr., O.E., 2016. Carcass type affects local scavenger guilds more than habitat connectivity. PLoS One 11 (2), e0147798.

Oreskes, N., 2004. The scientific consensus on climate change. Science 306 (5702), 1686-1686.

Ostfeld, R.S., Keesing, F., 2000. Pulsed resources and community dynamics of consumers in terrestrial ecosystems. Trends Ecol. Evol. 15 (6), 232–237.

Paquet, P.C., 1992. Prey use strategies of sympatric wolves and coyotes in Riding Mountain National Park, Manitoba. J. Mammal. 73 (2), 337−343.

Parmesan, C., 2006. Ecological and evolutionary responses to recent climate change. Ann. Rev. Ecol. Evol. System. 37, 637−669.

Pulliam, H.R., 1974. On the theory of optimal diets. Am. Nat. 108 (959), 59−74.

Reeves, N.M., 2009. Taphonomic effects of vulture scavenging. J. Forensic Sci. 54 (3), 523−528.

Ricketts, D.R., 2013. Scavenging effects and scattering patterns on porcine carcasses in Eastern Massachusetts. Unpublished Masters thesis. School of Medicine. Boston University, Boston.

Ripple, W., Beschta, R., 2012. Trophic cascades in Yellowstone: the first 15 years after wolf reintroduction. Biol. Conserv. 145 (1), 205−213.

Ripple, W., Larsen, E., Renkin, R., Smith, D., 2001. Trophic cascades among wolves, elk, and aspen in Yellowstone National Park's Northern Range. Biol. Conserv. 102 (3), 227−234.

Rippley, A., Larison, N.C., Moss, K.E., Kelly, J.D., Bytheway, J.A., 2012. Scavenging behavior of Lynx rufus on human remains during the winter months of southeast Texas. J. Forensic Sci. 57 (3), 699−705.

Root, R.B., 1967. The niche exploitation pattern of the blue-gray gnatcatcher. Ecol. Monogr. 37 (4), 317−350.

Sebastián-González, E., Moleón, M., Gibert, J.P., Botella, F., Mateo-Tomás, P., Olea, P.P., et al., 2016. Nested species-rich networks of scavenging vertebrates support high levels of interspecific competition. Ecology 97 (1), 95−105.

Sebastián-González, E., Sánchez-Zapata, J.A., Donázar, J.A., Selva, N., Cortés-Avizanda, A., Hiraldo, F., et al., 2013. Interactive effects of obligate scavengers and scavenger community richness on lagomorph carcass consumption patterns. Ibis 155 (4), 881−885.

Selva, N., Fortuna, M.A., 2007. The nested structure of a scavenger community. Proc. R. Soc. B Biol. Sci. 274 (1613), 1101−1108.

Selva, N., Jędrzejewska, B., Jędrzejewski, W., Wajrak, A., 2005. Factors affecting carcass use by a guild of scavengers in European temperate woodland. Can. J. Zool. 83 (12), 1590−1601.

Semenza, J.C., Hall, D.E., Wilson, D.J., Bontempo, B.D., Sailor, D.J., George, L.A., 2008. Public perception of climate change: voluntary mitigation and barriers to behavior change. Am. J. Prevent. Med. 35 (5), 479−487.

Simberloff, D., Dayan, T., 1991. The guild concept and the structure of ecological communities. Ann. Rev. Ecol. System. 22 (1), 115−143.

Smith, H.R., DeGraaf, R.M., Miller, R.S., 2002. Exhumation of food by turkey vulture. J. Raptor Res. 36 (2), 144−145.

Spradley, M.K., Hamilton, M.D., Giordano, A., 2012. Spatial patterning of vulture scavenged human remains. Forensic Sci. Int. 219, 57−63.

Stahler, D., Heinrich, B., Smith, D., 2002. Common ravens, Corvus corax, preferentially associate with grey wolves, Canis lupus, as a foraging strategy in winter. Anim. Behav. 64 (2), 283−290.

Stahler, D.R., Smith, D.W., Guernsey, D.S., 2006. Foraging and feeding ecology of the gray wolf (Canis lupus): lessons from Yellowstone National Park, Wyoming, USA. J. Nutr. 136 (7), 1923S−1926S.

Stephens, D.W., 2008. Decision ecology: foraging and the ecology of animal decision making. Cogn. Affect. Behav. Neurosci. 8 (4), 475−484.

Switalski, T.A., 2003. Coyote foraging ecology and vigilance in response to gray wolf reintroduction in Yellowstone National Park. Can. J. Zool. 81 (6), 985−993.

Tomberlin, J.K., Crippen, T.L., Tarone, A.M., Singh, B., Adams, K., Rezenom, Y.H., et al., 2012. Interkingdom responses of flies to bacteria mediated by fly physiology and bacterial quorum sensing. Anim. Behav. 84 (6), 1449–1456.

Vander Wall, S.B., 1990. Food-hoarding mammals. In: Vander Wall, S.B. (Ed.), Food Hoarding in Animals. Princeton University of Chicago Press, Chicago, pp. 217–282.

Wallace, M.P., Temple, S.A., 1987. Competitive interactions within and between species in a guild of avian scavengers. Auk 104, 290–295.

Walther, G.-R., Post, E., Convey, P., Menzel, A., Parmesan, C., Beebee, T.J., et al., 2002. Ecological responses to recent climate change. Nature 416 (6879), 389–395.

Weber, E.U., Stern, P.C., 2011. Public understanding of climate change in the United States. Am. Psychol. 66 (4), 315.

Wilmers, C.C., Getz, W.M., 2005. Gray wolves as climate change buffers in Yellowstone. PLoS Biol. 3 (4), e92.

Wilmers, C.C., Post, E., 2006. Predicting the influence of wolf-provided carrion on scavenger community dynamics under climate change scenarios. Global Change Biol. 12 (2), 403–409.

Wilmers, C.C., Stahler, D.R., Crabtree, R.L., Smith, D.W., Getz, W.M., 2003. Resource dispersion and consumer dominance: scavenging at wolf-and hunter-killed carcasses in greater Yellowstone, USA. Ecol. Lett. 6 (11), 996–1003.

Wilson, E.E., Wolkovich, E.M., 2011. Scavenging: how carnivores and carrion structure communities. Trends Ecol. Evol. 26 (3), 129–135.

Wright, T.F., Eberhard, J.R., Hobson, E.A., Avery, M.L., Russello, M.A., 2010. Behavioral flexibility and species invasions: the adaptive flexibility hypothesis. Ethol. Ecol. Evol. 22, 393–404.

Young, A., Stillman, R., Smith, M.J., Korstjens, A.M., 2014. An experimental study of vertebrate scavenging behavior in a Northwest European woodland context. J. Forensic Sci. 59 (5), 1333–1342.

CHAPTER 7

Adapting Your Investigation: Recovery and Interpretation

INTRODUCTION

THE IMPORTANCE OF ARCHAEOLOGY

FORENSIC ARCHAEOLOGY AND SCAVENGERS:
ADAPTING THE METHODS EMPLOYED

CONSIDERATIONS FOR POSTMORTEM INTERVAL AND TRAUMA

INVESTIGATING VERTEBRATE-SCAVENGED REMAINS:
BEST PRACTICES

CONCLUSION

REFERENCES

INTRODUCTION

Knowledge of scavenger behavior can have a significant impact on whether forensic investigations of vertebrate-scavenged remains are effective. The ecological variables discussed throughout this book will assist in the identification of scavengers, enabling informed forensic investigators to predict scavenger behavior. Knowledge of scavenging behavior also adds context critical to reconstruction of the postmortem period and can improve forensic investigation in two significant ways: (1) by maximizing recovery of remains, and (2) via improved interpretation of perimortem trauma and postmortem damage. However, such knowledge must be properly tapped, and the use of archaeological methods is essential to proper scene documentation, evidence recovery (including evidence of scavengers present), and overall interpretation.

This chapter will undertake a brief overview of the benefits of archaeological methods; however, a comprehensive treatment of all relevant methods is beyond the scope of this text. There are several excellent sources that focus exclusively on field methods, and the

Forensic Taphonomy and Ecology of North American Scavengers. DOI: https://doi.org/10.1016/B978-0-12-813243-2.00007-5

interested reader is encouraged to refer to those for more detailed information than can be provided here. In addition, several universities offer short courses in field search and recovery methods geared specifically toward law enforcement professionals or students. Following presentation of the benefits of using archaeological methods at outdoor death scenes, we offer several specific examples from the literature about how adapting a particular search and recovery strategy with scavenger presence in mind is beneficial. We conclude with an investigation strategy for outdoor scenes that incorporates scavenger behavior and ecology, and provide a theoretical example of how the strategy may be applied in practice.

THE IMPORTANCE OF ARCHAEOLOGY

Archaeology is the study of the human past. The past is defined broadly, to include the time period from yesterday to millions of years ago. Most archaeologists work in time periods beginning hundreds (historic) to thousands (prehistoric) or even millions of years ago (paleoanthropological, i.e., with human ancestors). However, regardless of the time period being targeted, archaeological methods remain more or less the same for gathering evidence of past human activity. This is true whether a crime scene from five years ago or a transient campsite from 5000 years ago is being investigated.

Dirkmaat et al. (2008) discuss the maturation of forensic anthropology since its official founding several decades ago. What began as an application of osteology to skeletons recovered by law enforcement primarily for purposes of identification (Stewart, 1979) has become a fully-fledged discipline in its own right that incorporates trauma analysis, taphonomy, and archaeological methods to the analyses of human remains (Dirkmaat et al., 2008). Where law enforcement used to box up bones found in the woods and bring them to a laboratory, anthropologists are increasingly involved in the recovery process and continue to encourage law enforcement to either involve qualified professionals at outdoor death scenes or to seek out relevant training opportunities in forensic archaeology.

Archaeological methods differ substantially from popular depictions of excavation or outdoor crime scene investigation in that they are essentially designed to record spatial relationships. These spatial relationships importantly include not only physical dimensions, but also a temporal dimension. For example, just because two objects are located physically near one another does not mean that they were deposited at the same time. The events leading to their individual depositions could be many years apart. The concepts that describe these spatial and temporal dimensions are: stratification, context, association, and provenience (Dirkmaat, 2012a; Renfrew and Bahn, 2004).

Geologically, sediments and other deposits are laid down sequentially through time. These deposits therefore become stratified or layered, and archaeologists are interested in stratigraphy (the processes leading to stratification) as it helps to understand the order in which strata were deposited, i.e., the temporal dimension (Dirkmaat, 2012a). Humans and other organisms can contribute to stratification via normal (or criminal) activity (e.g., trash dumps, house foundations, burying a body, bone caches, etc.).

Context is best thought of as the representation of events leading to an object's deposition at a particular place and time, evoked by its spatial and temporal location. Context is reconstructed from the remaining two concepts: association and provenience. Association demonstrates that two or more items are linked together by virtue of having been deposited at about the same time (Dirkmaat, 2012a) or for the same purpose, while provenience is the three-dimensional location of an object (Renfrew and Bahn, 2004). In practice, all these concepts would work together for a crime scene as follows: the location of body parts deposited in the woods and scattered about by canids and timing of when it occurred represents the *context* (human criminal activity followed by scavenger activity); the body itself is becoming incorporated into the *stratification* of the site; the human bones are *associated* with each other as well as with the canid tracks and fur left behind; and the physical location of any given bone in x, y, and z coordinates forms that bone's *provenience*.

The only way to completely understand context is through methods designed to capture provenience and association, i.e., an object's place spatially and temporally. There is often only one opportunity to get it right, as archaeology frequently employs destructive techniques. For example, in the case of a surface-scattered body, as soon as the various bones and any other evidence are collected, their association and provenience are lost forever. Photographs by themselves do not provide sufficient documentation to allow accurate context reconstruction. This reconstruction is of course important because it can provide information such as time since death, decedent identity, perpetrator behavior, and possible perimortem trauma, among other things (Dirkmaat, 2012b; Sorg et al., 2012).

Forensic anthropology has embraced archaeology as it solves problems related to outdoor death scene reconstruction. Unfortunately, law enforcement has been slow to catch up (Dirkmaat, 2012a). Dirkmaat (2012a) reviewed law enforcement literature relevant to crime scene processing and while he found a plethora of information related to indoor crime scenes, there was a dearth of information on outdoor scenes. Systematic archaeological techniques can resolve some of the inherent complexity in outdoor death scenes, and forensic anthropologists, given their frequent collaboration with law enforcement, are in a unique position to provide counsel on best approaches for any given outdoor scene.

Dupras et al. (2006), Holland and Connell (2009), Cheetham and Hanson (2009), and Dirkmaat (2012a) (among others) offer comprehensive treatments of archaeological methods applied to outdoor death scenes, and so we would not attempt to reiterate the details here. However, it is important to emphasize that documentation rounds out the top three best practices when it comes to archaeology: (1) documentation, (2) documentation, and (3) documentation. These include written records, maps to scale, and photographs (Dirkmaat, 2012a). Such a triad should ideally allow for the reconstruction of a site once it has been destroyed—in the case of an outdoor death scene, once the remains and any associated evidence have been recovered. It is important to remember that the only people who will ever physically see the outdoor scene before it is lost forever will be those on the search and recovery team. It is therefore their responsibility to record all the information at a comprehensive enough level so that the final interpretation can be understood and scrutinized by others not present, including family members, professionals in the medicolegal community, and the public.

FORENSIC ARCHAEOLOGY AND SCAVENGERS: ADAPTING THE METHODS EMPLOYED

Generally, there are several strategies that can be employed to maximize recovery of remains at outdoor death scenes. In archaeology, the old idiom applies: "there's more than one way to skin a cat." As long as the documentation is sufficient as outlined earlier, any number of strategies or methods that record association and provenience, and when relevant, stratigraphy, should suffice. The possible presence of scavengers complicate planned recovery strategies, but as Sorg et al. (2012: 480) highlight, "forensic taphonomy adopts an ecological perspective," underscoring the point being made throughout this book that knowledge about scavenger ecology will ultimately lead to improved skeletal analyses and improved contextual reconstructions, including time since death.

Haglund (1997) was the first to outline any specific search strategy with regard to recovering human remains while considering the impact of scavengers. Specifically, he was aiming to delineate how best to locate missing teeth from a decedent's body left as a surface deposit. He lists four questions, the answers to which should allow focus of employable methods. The questions are as follows:

1. Are the remains scattered?
2. From where was the body scattered?
3. What were the most likely trajectories of dispersion?
4. Are there any special circumstances that might affect disassociation of [remains] or scatter? (Haglund, 1997: 390).

When remains are absent or missing, the answers to these questions can assist with refocusing the investigation so that remains and associated evidence recovery are maximized. Animals are known to typically scatter remains along game trails (Haglund, 1997; Kjorlien et al., 2009) and unique characteristics of either the body (e.g., dismemberment by a perpetrator) or environment (e.g., difficult terrain) can affect where scavengers may have transported the remains (Haglund, 1997).

Following the tradition of experimental research in forensic taphonomy, VanLaerhoven and Hughes (2008) placed butchered pigs at two distinct sites: a wooded forest and a meadow grassland to identify (1) how scavengers would affect the remains in these different habitats, and (2) how quickly each site could be searched for the remains after several days, when four pedestrian search methods were employed. These methods were: (1) link, (2) line, (3) zone, and (4) spiral. Such search techniques are essentially variations of pedestrian surveys which can take any one of several patterns, commonly used in archaeology to perform initial assessment of sites (Holland and Connell, 2009). The link method uses spatial and environmental context of located evidence to determine the most probable path of movement of additional evidence. In the line method, searchers sweep over a search area while remaining parallel to one another. The zone, or grid, method divides a search area and assigns searchers or teams of searchers to each partitioned unit. Finally, searchers employing the spiral strategy walk the search area in increasingly large concentric circles from a given starting point. In their study, VanLaerhoven and Hughes (2008) were unable to compare the effectiveness of these methods due to the variation of scavengers (and thus variation in scattering patterns) that affected their sample. However, they did find that generally, forest habitats take significantly less search time than meadow habitats (VanLaerhoven and Hughes, 2008).

These findings demonstrate the importance of scavenger identification, and behavioral prediction, in selecting a search and recovery strategy. In recovery efforts of simulated fox-scavenged remains, Young et al. (2016) demonstrated that search teams were more than twice as likely to recover scattered skeletal elements when first provided with information on fox scattering patterns. To adapt their search and recovery strategies, the teams that were informed chose to employ the "Winthrop method," or reference point method, which concentrates search efforts at specific locations (or points) (Young et al., 2016). The specificity of the reference point method is particularly useful when remains have been scavenged by caching species. For example, reference points in a fox scavenging scenario include the bases of trees and dense vegetation, where foxes commonly cache prey (Macdonald, 1976; Young et al., 2016). These reference points may also be useful in recovering remains scavenged by other carnivores, as the propensity to relocate remains to areas of dense vegetation is reported in other canids, bears, and felids (Vander Wall, 1990). Relevant reference points in a scavenging scenario may also include canid dens, avian nests, rodent burrows, and game trails (Bankaitis, 2012; Haglund, 1997; Haglund et al., 1990).

Recovery strategy may also be adapted for other scavenging behaviors. For example, larger carnivores are capable of moving body parts, or even entire carcasses, from the original deposition site. When remains have been moved considerable distances, scattering typically occurs in a linear pattern away from the deposition location (Haglund, 1997; Jones, 2011). If the scattered remains are the primary discovery and the location of decomposition can be identified by isolated changes in soil or vegetation, focusing search efforts between the two points and continuing along this direction of movement using the link method may maximize recovery efforts (Haglund, 1997; Jones, 2011).

If the deposition location is the primary discovery, general knowledge of scattering patterns may help define the search area for additional remains. As discussed in the previous chapter, scavengers in more populated areas tend to scatter remains away from loci of human activity (Haglund, 1997; Kjorlien et al., 2009). Species-specific scattering patterns have also been identified. For example, Bright (2011) demonstrated that American black bears (*Ursus americanus*) in northern California tend to transport remains upslope, while a study of vulture (Cathartidae spp.) scavenging patterns in central Texas has demonstrated a gradual movement of skeletal elements downslope (Spradley et al., 2012).

CONSIDERATIONS FOR POSTMORTEM INTERVAL AND TRAUMA

Once remains have been appropriately collected and documented, an understanding of vertebrate scavenging behavior continues to play an important role in the interpretation of perimortem and postmortem events at a death scene. Vertebrate scavenging has variable effects on the estimation of postmortem interval (PMI). Deviation of the true PMI from the PMI estimated using total body scores (Megyesi et al., 2005) and accumulated degree days (Vass et al., 1992) has been noted for remains exposed to both avian and terrestrial scavengers (Smith, 2015; Suckling et al., 2016). Although vertebrate scavenging is known for its potential to accelerate decomposition, it should be noted that scavenging may skew PMI estimates by influencing other taphonomic agents (Smith, 2015). For example, vertebrate scavenging focused at natural orifices may disrupt insect colonization of remains by displacing egg or larval masses, delaying the progression of decomposition and leading to underestimations of the PMI (King et al., 2016; Pechal et al., 2014).

Being able to distinguish antemortem or perimortem trauma from postmortem damage is critical in death scene reconstruction. The condition of the remains should be documented in detail (including anatomical location and characteristics of defects/damage) to distinguish perimortem trauma from postmortem damage created by scavengers (invertebrate or vertebrate) or abiotic factors (Binford, 1981; Haglund and Sorg, 2001). Identifying characteristic postmortem damage caused by vertebrate scavenging was discussed in detail in previous chapters and will not be reiterated here. However, as a final point, knowledge of species-typical scavenging patterns may also result in the overgeneralization of any present defects as due to postmortem effects, without considering what existing trauma taphonomic processes may have obscured (Pokines and Symes, 2014). Overall, vertebrate scavenging complicates trauma analysis and any damage or anomalies on the remains requires additional scrutiny.

INVESTIGATING VERTEBRATE-SCAVENGED REMAINS: BEST PRACTICES

Investigation of vertebrate-scavenged remains must be conducted systematically and include consideration of ecological variables. A recommended approach for investigating such remains in the field is outlined below*:

1. Be aware of those scavenging species found in the geographic area, including wild, introduced, and domestic species.
2. Carry out recommended archaeological search and recovery strategies, as discussed in this chapter and elsewhere.
3. Collect faunal evidence from the scene, such as hair or fur, feathers, scat, and tracks. Hair or fur, feathers, and photographs of tracks can be identified to species-level by a local wildlife biologist or similar professional. Scat of larger carnivores (i.e., dog size or larger), when suspected to contain bone, can be processed by the anthropologist for the presence of bone fragments.
4. Inventory skeletal elements and record the characteristics and locations of damage.
5. Identify that vertebrate scavenging affected the remains, distinguishing postmortem animal scavenging damage from perimortem trauma and damage caused by other taphonomic processes.
6. Consider diagnostic damage patterns of candidate scavengers, summarized in Table 7.1 and detailed in Chapter 5.
7. Based on the above, identify candidate scavengers which are most likely to produce the observed damage or which have been linked to the scene by faunal evidence.

8. Consider ecological variables affecting the scene and additional variables (i.e., abiotic taphonomic agents or invertebrate scavengers) influencing the condition of the remains. Determine the most likely vertebrate scavenger(s) given the scene's context.
9. Given the environment and scavenger identification, predict scavenger behavior and adjust recovery strategy accordingly.
10. Use all recovered and interpreted information to estimate the postmortem interval.

*Note that the completion of all steps may require multiple trips to the scene, before recovery and after an anthropologist has analyzed the remains.

A theoretical example of how the preceding may occur is as follows: As a member of a local law enforcement's evidence response team located in the northeastern United States, you are made aware of a call to emergency services reporting the discovery of a human head. Upon arrival at the scene, you and your colleagues note that the location is a large wooded area with sporadic hunting cabins dispersed throughout. A large lake is within a quarter of a mile. The time of the year is summer. The head is in fact a skull, devoid of all tissue, resting against a log. Given the fact that you are located in the northeastern United States, you are aware that common scavengers for wooded areas include, but are not limited to domestic dogs, coyotes, black bears, bobcats, rats/mice, squirrels, white-tailed deer, and crows (*step 1*).

You and your colleagues call a local forensic anthropologist or archaeologist for assistance, who works with you to devise the best search strategy given the context (*step 2*). During the search, several tufts of animal fur and scat are noted in proximity to concentrations of human remains. Following documentation, all of the scat and some of the fur is collected for further analysis (*step 3*). A skeletal inventory is taken at the scene and again when the remains are transported to the forensic anthropology laboratory or morgue, and it is noted that one of the upper limbs is absent (*step 4*).

Following skeletal inventory, the anthropologist undertakes common procedures to analyze the remains, which includes an assessment of taphonomic damage and perimortem trauma. She identifies a perimortem gunshot wound to the cranium that exhibits parallel furrows along the edges of the entrance defect, as well as epiphyseal ends that

Table 7.1 Diagnostic Taphonomic Signatures

Taxon	Diagnostic Features	Scatter Potential[a]	Caching Behavior
Canidae Canids	• V-shaped punctures with stretch lacerations in soft tissue • Heavy gnawing, especially of epiphyses	High	Yes
Cathartidae Vultures	• Early disarticulation of the cranium and mandible • Frayed connective tissue • Stretching of skin around orifices and wounds • Bone marking skewed toward scratches on thick cortical areas (e.g., diaphyses)	High	No (with exceptions, i.e., condors)
Cervidae Deer	• Bone marking skewed toward furrows superimposed over dry bone • Fork-shaped defects at bone ends	Low	No
Corvidae Corvids	• Frayed connective tissue • Stretching of skin around orifices and wounds • Bone marking skewed toward shallow scratches	High	Yes
Crocodilia Crocodilians	• Amputation of appendages with shredded tissue • Bisected tooth marks and hook scoring, conspicuous lack of furrowing	High	No
Didelphidae Opossums	• Minimal external soft tissue damage • Splintered costal rib ends without crushing damage	Low	Yes
Felidae Felids	• Incised wound margins • Low density of tooth marks	Low	Yes
Procyonidae Raccoons	• Hollowed appendages with entry wounds at joints • High density of tooth marks in hands, feet, ribs, and vertebrae	Low	Unknown
Rodentia Rodents	• Rounded, crater-like destruction of soft tissue • Shallow, overlapping, paired furrows on bone	Low	Yes
Selachimorpha Sharks	• Amputation of appendages with incised edges • Parallel linear and spiral gouges with associated bone shaving	High	No
Suidae Swine	• Damage concentrated in axial skeleton • Bone marking skewed toward broad scores with flat bottoms • L-shaped punctures	Low	No
Ursidae Bears	• Heavy destruction of the axial skeleton • Scores have U-shaped cross section • Tooth marks associated with crushing damage	High	Yes

[a] *Refers to the scavenger's ability to move remains or disperse them over a wide area, including transport inside and eventual elimination from the digestive system.*

have multiple punctures and scoring marks along diaphyses (*step 5*). Such marks are characteristic of rodent gnawing and canid gnawing, respectively (*step 6*). The biologist's report indicates that the fur found at the scene is coyote fur, positively putting coyotes in association with the remains (*step 7*).

Following these findings, discussions among the team members (law enforcement, biologist, anthropologist, and pathologist) are undertaken. Conversation focuses around the ecological context (northeastern U.S. forest that is relatively devoid of humans, summertime, local fresh water source). A number of beetles were present under the remains of the torso, and blowfly pupal casings were present inside the skull. Soil staining of the bones was also present, although some bones had been bleached by the sun (*step 8*).

Given the outcome of the conversation and/or interpretation that took place during step 8, the team decides to go back to the scene to search for the missing arm. Coyotes have already been identified as one of the vertebrate scavengers affecting the remains, from the skeletal marks as well as from faunal evidence at the scene. All canids are known cachers of food, and they will also bring food to dens for whelping pups and nursing mothers. A possible den had been identified during the first search, and during the second search, it is examined more closely, at which point human arm bones are discovered in the leaf litter just outside the den (*step 9*).

The anthropologist evaluates the state of the remains to include the level of decomposition (fully skeletal), the scavenging damage present by coyotes and rodents, the evidence of invertebrate colonization of the remains, and the evidence of abiotic effects (i.e., soil staining and sun bleaching); as well as the evidence that coyotes removed an upper limb to feed to pups. Recall that the time of year is summer and coyote pups are whelped in spring. However, these remains are fully skeletal with evidence of sun bleaching and rodent damage, two indicators that typically do not occur until later in the decomposition process. This is also the northeastern United States, where winters are long and cold and spring comes relatively late. These factors, taken together, indicate that the individual has likely been deceased since *at least* the late winter/early spring of the year before, rather than of the current year (*step 10*).

The preceding example demonstrates how methods and knowledge from various areas (archaeology, forensic anthropology, ecology, and law enforcement) can be used in concert to affect a successful field recovery and laboratory analysis. Consider an alternative scenario, where the team goes out to the scene without a plan, assistance from an expert, or knowledge of archaeology or ecology. The likely result is a relatively disorganized wandering around the woods, randomly picking up suspected bones and boxing them up. In this scenario, not only is bone recovery minimized, but the loss of contextual information makes it difficult if not impossible for event reconstruction later.

CONCLUSION

The good news is that archaeological techniques and knowledge of scavenger ecology can be incorporated by law enforcement without too much effort. Agencies may benefit from sending representatives to participate in one of many short courses offered by several universities that focus on recovery techniques. Further, law enforcement should seriously consider partnering with a local forensic archaeologist or board-certified forensic anthropologist who could provide assistance when the need arises. Such professionals are highly trained and have the requisite experience and expertise to be of service.

REFERENCES

Bankaitis, J., 2012. Examination of scavenging associated with wolves. Unpublished Masters thesis. Department of Anthropology. The University of Montana, Missoula.

Binford, L.R., 1981. Patterns of bone modifications produced by nonhuman agents. In: Binford, L.R. (Ed.), Bones: Ancient Men and Modern Myths. Academic Press, San Diego, pp. 35–86.

Bright, L.N., 2011. Taphonomic signatures of animal scavenging in northern California: a forensic anthropological analysis. Unpublished Masters thesis. Department of Anthropology. California State University, Chico.

Cheetham, P., Hanson, I., 2009. Excavation and recovery in forensic archaeological investigations. In: Blau, S., Ubelaker, D. (Eds.), Handbook of Forensic Anthropology and Archaeology. Left Coast Press, Walnut Creek, pp. 141–150.

Dirkmaat, D., 2012a. Documenting context at the outdoor crime scene: why bother? In: Dirkmaat, D. (Ed.), A Companion to Forensic Anthropology. Wiley-Blackwell, Malden, pp. 48–65.

Dirkmaat, D., 2012b. Introduction to Part II. In: Dirkmaat, D. (Ed.), A Companion to Forensic Anthropology. Wiley-Blackwell, Malden, pp. 43–47.

Dirkmaat, D.C., Cabo, L.L., Ousley, S.D., Symes, S.A., 2008. New perspectives in forensic anthropology. Am. J. Phys. Anthropol. 137 (S47), 33–52.

Dupras, T., Schultz, J., Wheeler, S., Williams, L., 2006. Forensic Recovery of Human Remains: Archaeological Approaches. CRC Press, Boca Raton.

Haglund, W., 1997. Scattered skeletal human remains: search strategy considerations for locating missing teeth. In: Haglund, W.D., Sorg, M.H. (Eds.), Forensic Taphonomy: The Postmortem Fate of Human Remains. CRC Press, Boca Raton, pp. 383–394.

Haglund, W.D., Reichert, D.G., Reay, D.T., 1990. Recovery of decomposed and skeletal human remains in the 'Green River Murder' investigation. Implications for medical examiner/coroner and police. Am. J. Forensic Med. Pathol. 11 (1), 35–43.

Haglund, W.D., Sorg, M.H. (Eds.), 2001. Advances in Forensic Taphonomy: Method, Theory, and Archaeological Perspectives. CRC Press, Boca Raton.

Holland, T., Connell, S., 2009. The search for and detection of human remains. In: Blau, S., Ubelaker, D. (Eds.), Handbook of Forensic Anthropology and Archaeology. Left Coast Press, Walnut Creek, pp. 129–140.

Jones, A.L., 2011. Animal scavengers as agents of decomposition: the postmortem succession of Louisiana wildlife. Unpublished Masters thesis. Department of Geography and Anthropology. Louisiana State University, Baton Rouge.

King, K.A., Lord, W.D., Ketchum, H.R., O'Brien, R.C., 2016. Postmortem scavenging by the Virginia opossum (*Didelphis virginiana*): impact on taphonomic assemblages and progression. Forensic Sci. Int. 266 (576), e1-576.e6.

Kjorlien, Y.P., Beattie, O.B., Peterson, A.E., 2009. Scavenging activity can produce predictable patterns in surface skeletal remains scattering: observations and comments from two experiments. Forensic Sci. Int. 188, 103–106.

Macdonald, D.W., 1976. Food caching by red foxes and some other carnivores. Z. Tierpsychol. 42 (2), 170–185.

Megyesi, M.S., Nawrocki, S.P., Haskell, N.H., 2005. Using accumulated degree-days to estimate the postmortem interval from decomposed human remains. J. Forensic Sci. 50 (3), 618–626.

Pechal, J.L., Benbow, M.E., Crippen, T.L., Tarone, A.M., Tomberlin, J.K., 2014. Delayed insect access alters carrion decomposition and necrophagous insect community assembly. Ecosphere 5 (4), 1–21.

Pokines, J., Symes, S.A. (Eds.), 2014. Manual of Forensic Taphonomy. CRC Press, Boca Raton.

Renfrew, C., Bahn, P., 2004. Archaeology: Theories, Methods and Practice. Thames & Hudson, New York.

Smith, J.K., 2015. Raccoon scavenging and the taphonomic effects on early human decomposition and PMI estimation. Unpublished Masters thesis. Department of Anthropology. University of Tennessee, Knoxville.

Sorg, M., Haglund, W., Wren, J., 2012. Current research in forensic taphonomy. In: Dirkmaat, D. (Ed.), A Companion to Forensic Anthropology. Wiley-Blackwell, Malden, pp. 477–498.

Spradley, M.K., Hamilton, M.D., Giordano, A., 2012. Spatial patterning of vulture scavenged human remains. Forensic Sci. Int. 219, 57–63.

Stewart, T.D., 1979. Essentials of Forensic Anthropology: Especially as Developed in the United States. Charles C. Thomas, Springfield.

Suckling, J.K., Spradley, M.K., Godde, K., 2016. A longitudinal study on human outdoor decomposition in central Texas. J. Forensic Sci. 61 (1), 19–25.

Vander Wall, S.B., 1990. Food-hoarding mammals. In: Vander Wall, S.B. (Ed.), Food Hoarding in Animals. Princeton University of Chicago Press, Chicago, pp. 217–282.

VanLaerhoven, S.L., Hughes, C., 2008. Testing different search methods for recovering scattered and scavenged remains. Can. Soc. Forensic Sci. J. 41 (4), 209–213.

Vass, A.A., Bass, W.M., Wolt, J.D., Foss, J.E., Ammons, J.T., 1992. Time since death determinations of human cadavers using soil solution. J. Forensic Sci. 37 (5), 1236–1253.

Young, A., Stillman, R., Smith, M.J., Korstjens, A.H., 2016. Applying knowledge of species-typical scavenging behavior to the search and recovery of mammalian skeletal remains. J. Forensic Sci. 61 (2), 458–466.

CHAPTER 8

Suggestions for Future Directions

INTRODUCTION
MISSING DATA AND NEW QUESTIONS
STUDYING SCAVENGERS: CHALLENGES AND RESEARCH DESIGN
BRINGING IT ALL TOGETHER: KNOWLEDGE DISTRIBUTION
CONCLUSION
REFERENCES

INTRODUCTION

Vertebrate scavenging behavior is an area that is ripe for further research. Gaps persist on the taphonomic effects of certain taxa generally as well as in specific ecological contexts. While basic patterns have been identified, more complicated questions have emerged regarding how to best address the effects that vertebrate scavenging can have on forensically-important questions that surround identification, circumstances of death, and postmortem interval (PMI). This concluding chapter presents an overview of topics that are missing from the literature or warrant additional investigation, discusses approaches that lend themselves to studying scavenging behavior within an ecological framework, and pushes for further development of standards for investigating scavenged death scenes.

MISSING DATA AND NEW QUESTIONS

There are several common scavenging species that have yet to have their taphonomic signatures described in detail, despite frequent reports of visits to monitored carcasses (Table 8.1). For example, the Virginia opossum (*Didelphis virginiana*, Fig. 5.34) and striped skunk (*Mephitis mephitis*, Fig. 8.1) have been identified as scavengers in numerous taphonomic studies (Allen et al., 2015; Bauer et al., 2005;

Forensic Taphonomy and Ecology of North American Scavengers. DOI: https://doi.org/10.1016/B978-0-12-813243-2.00008-7

Table 8.1 Understudied Scavenging Species of North America			
Species	Conservation Status	Habitat	Possible Forensic Impacts
Red-Tailed Hawk[a] (*Buteo jamaicensis*)	Least concern	Open fields, ubiquitous throughout continent	-Soft tissue damage -Bone modification
Crested Caracara[a] (*Caracara cheriway*)	Least concern	Open fields, restricted to southernmost United States and Central America	-Soft tissue damage -Bone modification
North American Porcupine[b] (*Erethizon dorsatum*)	Least concern	Deciduous and coniferous forest, brush	-Bone modification -Caching
Desert Tortoise[c] (*Gopherus agassizii*)	Vulnerable	Mojave and Sonoran Deserts of southwestern United States, northwest Mexico	-Bone modification
Wolverine[b] (*Gulo gulo*)	Least concern	Tundra, taiga (swampy coniferous forest), mountainous regions	-Soft tissue damage -Bone modification -Scattering -Caching
American Marten[b] (*Martes americana*)	Least concern	Coniferous or mixed forest, northern United States and Canada	-Soft tissue damage -Bone modification -Scattering -Caching
Fisher[b] (*Martes pennanti*)	Least concern	Coniferous or mixed forest, northern United States and Canada	-Soft tissue damage -Bone modification -Scattering -Caching
Striped Skunk[b] (*Mephitis mephitis*)	Least concern	Brush, open woods, deserts, and suburbs or farmland	-Soft tissue damage -Bone modification
American Badger[b] (*Taxidea taxus*)	Least concern	Open habitats including plains, prairies, meadows, and deserts	-Soft tissue damage -Bone modification -Scattering -Caching
Nile Monitor[d] (*Varanus niloticus*)	Unassessed	Isolated populations in Florida especially in mangrove swamps and marshes	-Soft tissue damage -Bone modification
Conservation statuses courtesy of IUCN Red List of Threatened Species. [a]*Cornell Lab of Ornithology (2015).* [b]*Reid (2006).* [c]*Walde et al. (2007).* [d]*Enge et al. (2004).*			

Jones, 2011; King et al., 2016; Morton and Lord, 2006; Moss, 2012; Olson et al., 2012; Reeves, 2009; Ricketts, 2013; Steadman et al., 2016). Despite this, the taphonomic effects of scavenging by the Virginia opossum have only recently been described by King et al. (2016), following a fortuitous study that recorded a preponderance of opossum scavengers on pig carcasses.

*Figure 8.1 Skunk (*Mephitis mephitis*). "Striped skunk in leaves" is in the public domain.*

In addition, researchers interested in the taphonomic effects of skunk and other mustelid species, including the American marten (*Martes americana*, Fig. 8.2), American badger (*Taxidea taxus*, Fig. 8.3), fisher (*Martes pennanti*, Fig. 8.4), and wolverine (*Gulo gulo*, Fig. 8.5), have not yet had much luck recording their taphonomic signatures despite the mustelid taxon's frequent use of large carrion (Huner and Peter, 2012; Mattisson et al., 2017). In a study of vertebrate scavengers in Massachusetts, fishers were reported in association with pig carcasses, observed to interact with equipment and were recorded following the scent of pig carcasses that had been displaced by coyotes (Ricketts, 2013). Although the fishers in this study were not recorded scavenging, their persistent presence throughout the study coupled with reports of mustelid scavenging in other regions (see following paragraph) suggests that the forensic implications of mustelid scavenging warrants additional research.

In addition to scavenging by mustelids, caching behavior by species in this family is also of interest, with observations of fishers and badgers caching large mammal carcasses in situ. For example, Huner and Peter (2012) document a fisher in Ontario, Canada feeding from and caching an American black bear (*Ursus americanus*) carcass by covering it with local soil, grasses and pine needles, and twigs; even using bones from a moose (*Alces alces*). In contrast, American badgers in the Great Basin Desert, Utah, were recently observed feeding from and burying the carcasses of juvenile cows (*Bos taurus*) on two separate occasions (Frehner et al., 2017). To cache the calves, a badger tunneled underneath the carcass until the soil collapsed, at which point the

*Figure 8.2 American marten (*Martes americana*). "Pine Marten" by Steve Slocomb is licensed under CC BY 2.0.*

*Figure 8.3 American badger (*Taxidea taxus*). "Badger" is in the public domain.*

animal scraped the surrounding soil over the carcass to backfill the hole (Frehner et al., 2017). In contrast, along the Alberta–British Columbia border in Canada, wolverines have been documented to scatter-hoard the bones and hides of wolf-killed moose, burying them

*Figure 8.4 Fisher (*Martes pennanti*). "Fisher 3" by Pacific Southwest Region USFWS is licensed under CC BY 2.0.*

*Figure 8.5 Wolverine (*Gulo gulo*). "Wolverine" by O. Peters is in the public domain.*

under snow and earth (Wright and Ernst, 2004). The authors of this study also noted that wolverines preferentially chose to cache at sites with good horizontal visibility, rather than in areas of dense forest, and that they preferentially traveled using well-worn trails forged by human or animal traffic (Wright and Ernst, 2004).

Other species are absent from the literature. For example, the North American porcupine (*Erethizon dorsatum*), a large rodent, has received little mention in the scavenging literature. This is despite this species' known propensity to gnaw dry bone, and in fact bone accumulation is well documented in their Old World counterparts (Pokines, 2014). Desert tortoise (*Gopherus agassizii*) have also been known to break off and consume small bits of bone, especially from conspecific carcasses (Walde et al., 2007). Nile monitors (*Varanus niloticus*), a large species of West African lizard introduced to southern Florida in the 1990s, are also known to scavenge carrion (Enge et al., 2004). Each species described earlier is known or suspected to interact with carrion, although additional research is needed to assess their potential forensic impacts.

Further, the almost conspicuous absence of some species from the forensic literature may be due in part to the nested structure of scavenger communities, as discussed earlier in this volume. Many animals typically scavenge from remains that are also being utilized by other vertebrate species. Determining which species contributed to postmortem damage is complicated by the obliteration or alteration of marks produced by smaller scavengers when larger, more voracious scavengers also interact with remains. For example, scavenging by birds of prey is frequently reported in general studies of scavenging guilds. In forensic contexts, red-tailed hawks (*Buteo jamaicensis*) and crested caracaras (*Caracara cheriway*) are frequently reported in association with human remains at the Forensic Anthropology Research Facility at Texas State University, although evidence of possible scavenging is lost amongst the damage produced by the wakes of American black vultures (*Coragyps atratus*) that typically follow (Reeves, 2009; Spradley et al., 2012).

In some cases, scavenger behavior has been well documented in a single ecological context, but has yet to be explored in other locations. For example, almost all of the forensic literature on scavenging by raccoons (*Procyon lotor*) has been published based on observations made at the Anthropology Research Facility in Knoxville, Tennessee (Jeong et al., 2016; Smith, 2015; Steadman et al., 2016; Synstelien, 2013; Synstelien and Klippel, 2005); with the exception of a single study conducted in Maine (Hannigan, 2015). Likewise, virtually all forensic research on vulture scavenging of human remains has been conducted at the Forensic Anthropology Center at Texas State University, located in south-central Texas (Klein, 2013; Reeves, 2009; Spradley et al., 2012), although studies using pigs as human analogs have been conducted in Arizona (Beck et al., 2015) and southern Illinois (Dabbs and Martin, 2013). Research on the forensic implications of vertebrate scavenging is particularly deficient in northern contexts, in large part due to the absence of human decomposition research facilities in the north, although researchers have begun to document scavenger interest in animal carcasses deposited during the winter in Maine (Sorg, 2011; Sorg et al., 2012b).

Conversely, while data on the behavior of several scavenging vertebrate species of interest are missing, scavenging behavior and the taphonomic effects produced have been thoroughly documented for many taxa. For example, in the forensic literature, canids (Haglund et al., 1989; Willey and Snyder, 1989) and rodents (Haglund, 1992) have received significant attention for decades. However, even for such well-documented species, questions pertaining to the application of scavenger knowledge and effects of cadaver-intrinsic and environmental variation persist. For example:

- How does the body composition of human remains (i.e., under or overweight) influence rates and patterns of soft tissue consumption?
- How does consumption pattern, soft tissue destruction, and bone modification compare between human remains and nonhuman animals?
- Does scavenging by particular taxa or particular scavenging guilds have predictable effects on PMI, and if so, how can PMI estimates be adjusted to compensate for scavenging damage?
- How does vertebrate scavenging impact bacterial succession or entomological activity, given that both lines of evidence are increasingly being used to produce PMI estimates?
- How do patterns of disarticulation and scattering compare between human adult and subadult (i.e., child) remains?
- How is the pattern of scattering affected by habitat characteristics, such as presence/absence of tree cover or proximity to bodies of water?

These questions, among others, remain pertinent to forensic casework and are viable avenues for future research endeavors. The following section will introduce research approaches that are often applied in studies of scavenger behavior along with several benefits and challenges of each.

STUDYING SCAVENGERS: CHALLENGES AND RESEARCH DESIGN

Studies of vertebrate scavenging are complicated from both theoretical and practical perspectives. Theoretically, animal behavior is governed and impacted by an overwhelming number of interacting variables; accounting for all of them is impossible. Practically, thorough observations of vertebrate scavengers cannot be conducted in a completely natural setting, as the mere presence of humans or introduction of novel monitoring devices (i.e., game cameras) can influence behavior. A notable example of this comes from a study of animal scavenging conducted in Massachusetts, where a field pole painted with stripes of a defined scale (to assist with assessment of scavenger size in photographs) was introduced (Ricketts, 2013). Fishers, known for their utilization of carrion, were documented investigating these poles extensively but showed little interest in the pig carcasses themselves, placed only a few meters away (Ricketts, 2013).

Studies of vertebrate scavenging are often conducted using one of two research designs: (1) actualistic or (2) experimental. An actualistic approach reports observations of natural phenomena (e.g., a forensic case report documenting postmortem damage produced by a domestic dog, or documentation of observations of scavenged carcasses found in the woods), while experimental studies are designed to test hypotheses (e.g., documentation of damage to pig femora fed to captive caimans (*Caiman crocodilus*) to assess consistency with the crocodilian bone modification pattern) (Sorg et al., 2012a). While actualistic approaches provide a more accurate representation of the true spectrum of variation possible, they offer little control over confounding variables. In contrast, experimental studies allow for more precise regulation of variables, but produce results that may not always accurately reflect what occurs in real-world scenarios.

Many ecological studies of scavenger behavior take an actualistic approach. Some have used GPS collars to track animals in real time, using spatial patterns in animal movement (i.e., repeated visits to specific locations) to locate animal kills, scavenged carcasses, or caches after the animals have abandoned the remains (Elbroch and Wittmer, 2013; Metz et al., 2011). These types of studies minimize human influences on animal behavior. However, the damage observed on an abandoned carcass only represents the final product of a scavenger's activity, providing no data about the condition of the carcass throughout the scavenging process. Additionally, in the absence of clear evidence otherwise (e.g., rodent gnawing present on a puma kill), actualistic studies may assume that the primary species (i.e., the specific one being tracked) was the only one to interact with the carcass.

Although actualistic ecological studies are limited in the kinds of data produced, they are very helpful in answering specific kinds of research questions because they provide information about undisturbed animal behavior in natural settings. Case reports of scavenged remains are considered to be actualistic studies, presenting data collected from remains that were scavenged. Although interpretations of scavenger-inflicted damage in case reports are not as precise or tidy as those made from experimental data, the present postmortem damage is obviously a more relevant representation of the taphonomic effects seen in actual forensic casework. Actualistic studies may be a fitting approach for addressing questions of animal behavior or general patterning of taphonomic effects, at least for those taxa with well-documented taphonomic signatures. For example, parallel scoring and furrowing can readily be attributed to rodent gnawing. Case reports of rodent-scavenged human remains may therefore be used to identify patterns in the distributions of these gnaw marks on the skeleton under natural conditions.

Experimental studies where carcasses or carcass parts are fed to captive animals are an alternative form of research useful for isolating the taphonomic signatures of scavenging species. This is because they control for the activity of other scavengers. Such studies have been conducted to analyze taphonomic signatures of large felids, canids, and alligators (Drumheller and Brochu, 2014; Pickering and Carlson, 2004; Willey and Snyder, 1989). However, extrapolating interpretations based on data gathered from captive animals to wild ones should be done with caution given behavioral differences, and thus differences in taphonomic effects, that may arise in captivity (Young et al., 2015a, b). Tooth marking and bone modification may be similar between captive and wild animals, while patterns of scatter and caching in captive settings are more likely to be influenced by variables such as the artificial environment, high (or low) density of conspecifics within the enclosure, or the continuous presence of humans.

Experimental studies may also be influenced by the differential use of human cadavers or nonhuman carcasses. Differences in carcass utilization have been demonstrated by Steadman et al. (2016), who observed that the consumption patterns of raccoons differed between human remains and pig carcasses. The distinction reflects differences in muscle distribution between pigs and humans; recall that raccoons have a preference for muscle tissue. Scavenging pattern differences between human cadavers and nonhuman analogs have also been demonstrated in contrasting studies of vultures (Dabbs and Martin, 2013; Reeves, 2009; Spradley et al., 2012); canids (Haglund, 1997; Pokines, 2014; Willey and Snyder, 1989); felids (Pickering, 2001; Rippley et al., 2012; Virchow and Hogeland, 1994); and other studies of raccoons (Hannigan, 2015; Jeong et al., 2016). However, the differences between these studies cannot be isolated by the species of the carcass used alone, given seasonal variation and intrinsic carcass qualities (e.g., body mass or absence of viscera due to autopsy, as in Rippley et al., 2012).

Natural experiments, such as those conducted at decomposition research facilities, seem to walk the line between actualistic and experimental studies. While certain variables can be manipulated, the controlled variables are largely limited to intrinsic characteristics of the deposited remains themselves or ecological characteristics of the deposition environment. For example, researchers could assess how patterns of carcass consumption are affected when decedents are dressed in heavy winter gear compared to light clothing, or conversely what scattering patterns exist for wooded versus grassland ecosystems. Other environmental characteristics, such as weather patterns, cannot be controlled but can be quantified and accounted for in later analyses. Similar to actualistic approaches, natural experiments aim to minimize disturbance of a scavenger's natural interaction with the carcass once placed.

Natural experiments may also be affected by technological limitations. For example, utilization of game cameras to monitor scavenging is common. These cameras are often programmed to take photographs when triggered by a motion sensor or after a given interval (e.g., once per hour). However, the sensitivity of motion sensors may not be sufficient to capture the activity of smaller scavengers (King et al., 2016), while using an intermittent setting may miss active scavenging that occurs between photographs. To maximize collection of data, King et al. (2016) recommend coupling game cameras with digital video recording systems to ensure that all scavenging species are documented. Game cameras and other recording systems should also be checked and repositioned regularly to verify that remains have not been dragged out of the frame.

BRINGING IT ALL TOGETHER: KNOWLEDGE DISTRIBUTION

Information that can be extracted from a death scene is limited by the knowledge of the investigators and the methods available for recovery and analysis. As discussed in the previous chapter, the efficiency and accuracy of investigative methods can be improved by knowledge and application of forensic research. As Haglund and Sorg (2001) observe, multidisciplinary training for investigators is useful for the analysis of taphonomic data, but is also time-consuming and expensive. In an ideal world, experts would be available at every death scene to assist with recovery and interpretation of remains in situ, but in reality law enforcement agencies must work within practical (i.e., financial and temporal) limits. Given the multidisciplinary nature of vertebrate scavenging studies, such information—if published—may be available from a wide variety of academic journals in different scholarly disciplines, and it is unreasonable to expect recovery teams to keep up with all of it on top of fulfilling their day-to-day duties.

This text has compiled information from case reports and academic research published over the last several decades into a single, accessible source to promote a medicolegal understanding of vertebrate scavenging that can be applied to forensic casework. However, while textbooks and field manuals have their place, there are more active ways to keep medicolegal professionals up-to-date on scavenger research. Anthropologists and other taphonomic researchers can educate law enforcement by publishing the results of their research in law enforcement newsletters, presenting at law enforcement conferences, or by offering workshops on strategic field recovery and excavation that are catered to professionals. Apart from educational efforts, knowledge could be perpetuated via a database of scavenged casework, which could make vertebrate scavenging data readily accessible to medicolegal professionals while simultaneously serving as a springboard for additional research initiatives. A well-designed, carefully managed database has the potential to reveal previously unrecognized patterns that link ecology, animal behavior, and taphonomic effects.

Outside the realm of research, an organized scavenging database could strengthen the behavioral predictions that promote accuracy in the reconstruction and interpretation of vertebrate-scavenged death scenes. Medicolegal professionals could reference the database to devise strategies to recover missing elements, in formulating their interpretation of trauma and scavenging damage, or adjusting PMI estimates. Furthermore, the use of a standardized scavenging data set could be drawn upon to lend credibility to an investigator's interpretations that are presented in court. For example, a body may be recovered from the Floridian coast with a postmortem amputation of the left forearm and near amputation of the other appendages. At autopsy, the medical examiner would conclude that the amputation was caused by a shark, rather than attempted dismemberment by a bungling murderer, due to the parallel, curved gouges present in the bones adjacent to the amputation sites. This finding could be considered more reliable if the damage could be likened to numerous confirmed and published cases of shark scavenging in the region, as opposed to relying on the medical examiner's own experience. This would serve to increase the likelihood that the interpretation could be admitted as evidence in court, should it be necessary.

CONCLUSION

This book is the first of its kind, summarizing data from ecological and forensic research and presenting it in a holistic format, easily accessible to students and seasoned investigators alike. It provides an overview of strategies for identifying responsible scavenging taxa, presents the basic taphonomic signatures of common North American vertebrate scavengers, and gives examples of how the ultimate condition of scavenged remains may be altered by ecological variables. Notably, this book lays out a recommended procedure for investigating scavenged outdoor death scenes within their ecological context.

Scavenged scenes are extremely disorganized and may present significant obstacles to sufficiently answer forensic questions and, in some cases, may affect the legal resolution of a case. The procedure and recommendations presented in this volume are guided by evolutionary and ecological theory and are designed to assist medicolegal professionals with maximizing recovery of remains in situ; while emphasizing documentation of scene context, thereby promoting accurate forensic interpretation of scavenger-induced damage. By delineating a theoretical basis and standardized approach for forensic analyses of vertebrate-scavenged remains, this volume offers a means to improve the scientific validity and reliability of the approaches used in such cases.

REFERENCES

Allen, M.L., Elbroch, L.M., Wilmers, C.C., Wittmer, H.U., 2015. The comparative effects of large carnivores on the acquisition of carrion by scavengers. Am. Nat. 185 (6), 822–833.

Bauer, J.W., Logan, K.A., Sweanor, L.L., Boyce, W.M., 2005. Scavenging behavior in puma. Southw. Nat. 50 (4), 466–471.

Beck, J., Ostericher, I., Sollish, G., De Leon, J., 2015. Animal scavenging and scattering and the implications for documenting the deaths of undocumented border crossers in the Sonoran Desert. J. Forensic Sci. 60 (S1), S11–S20.

Cornell Lab of Ornithology, 2015. All About Birds. Cornell University. Available from . Available from: https://www.allaboutbirds.org/, accessed 31.07.17.

Dabbs, G.R., Martin, D.C., 2013. Geographic variation in the taphonomic effect of vulture scavenging: the case for southern Illinois. J. Forensic Sci. 58 (S1), S20–S25.

Drumheller, S.K., Brochu, C.A., 2014. A diagnosis of *Alligator mississippiensis* bite marks with comparisons to existing crocodylian datasets. Ichnos 21 (2), 131–146.

Elbroch, L.M., Wittmer, H.U., 2013. Nuisance ecology: do scavenging condors exact foraging costs on pumas in Patagonia? PLoS One 8 (1), 1–8.

Enge, K.M., Krysko, K.L., Hankins, K.R., Campbell, T.S., King, F.W., 2004. Status of the Nile monitor (*Varanus niloticus*) in southwestern Florida. Southe. Nat. 3 (4), 571–582.

Frehner, E.H., Buechley, E.R., Christensen, T., Şekercioğlu, Ç.H., 2017. Subterranean caching of domestic cow (*Bos taurus*) carcasses by American badgers (*Taxidea taxus*) in the Great Basin desert, Utah. West. North Am. Nat. 77 (1), 124–129.

Haglund, W.D., 1992. Contribution of rodents to postmortem artifacts of bone and soft tissue. J. Forensic Sci. 37 (6), 1459–1465.

Haglund, W.D., 1997. Dogs and coyotes: postmortem involvement with human remains. In: Haglund, W.D., Sorg, M.H. (Eds.), Forensic Taphonomy: The Postmortem Fate of Human Remains. CRC Press, Boca Raton, pp. 367–382. Boca Raton.

Haglund, W.D., Reay, D.T., Swindler, D.R., 1989. Canid scavenging/disarticulation sequence of human remains in the Pacific Northwest. J. Forensic Sci. 34 (3), 587–606.

Haglund, W.D., Sorg, M.H., 2001. (Eds.), Advances in Forensic Taphonomy: Method, Theory, and Archaeological Perspectives.. CRC Press, Boca Raton.

Hannigan, A., 2015. A descriptive study of forensic implications of raccoon scavenging in Maine. Unpublished Honors thesis. Department of Anthropology. University of Maine, Orono.

Huner, E.A., Peter, J.F.B., 2012. In situ caching of a large mammal carcass by a fisher, Martes pennanti. Can. Field Nat. 126 (3), 234–237.

Jeong, Y., Jantz, L.M., Smith, J., 2016. Investigation into seasonal scavenging patterns of raccoons on human decomposition. J. Forensic Sci. 61 (2), 467–471.

Jones, A.L., 2011. Animal scavengers as agents of decomposition: the postmortem succession of Louisiana wildlife. Unpublished Masters thesis. Department of Geography and Anthropology. Louisiana State University, Baton Rouge.

King, K.A., Lord, W.D., Ketchum, H.R., O'Brien, R.C., 2016. Postmortem scavenging by the Virginia opossum (*Didelphis virginiana*): impact on taphonomic assemblages and progression. Forensic Sci. Int. 266 (576), e1-576.e6.

Klein, A.A., 2013. Vulture scavenging of pig remains at varying grave depths. Unpublished Masters thesis. Department of Anthropology. Texas State University, San Marcos.

Mattisson, J., Rauset, G.R., Odden, J., Andrén, H., Linnell, J.D.C., Persson, J., 2017. Predation or scavenging? Prey body condition influences decision-making in a facultative predator, the wolverine. Bull. Ecol. Soc. Am. 98 (1), 40–46.

Metz, M.C., Vucetich, J.A., Smith, D.W., Stahler, D.R., Peterson, R.O., 2011. Effect of sociality and season on gray wolf (*Canis lupus*) foraging behavior: implications for estimating summer kill rate. PLoS One 6 (3), e17332.

Morton, R.J., Lord, W.D., 2006. Taphonomy of child-sized remains: a study of scattering and scavenging in Virginia, USA. J. Forensic Sci. 51 (3), 475−479.

Moss, K.E., 2012. The effects of avian and terrestrial scavenger activity on human remains and decomposition in southeast Texas during an 18 month study. Unpublished Masters thesis. Department of Anthropology. University of Houston, Houston.

Olson, Z., Beasley, J., DeVault, T.L., Rhodes, O., 2012. Scavenger community response to the removal of a dominant scavenger. Oikos 121 (1), 77−84.

Pickering, T.R., 2001. Carnivore voiding: a taphonomic process with the potential for the deposition of forensic evidence. J. Forensic Sci. 46 (2), 406−411.

Pickering, T.R., Carlson, K.J., 2004. Baboon taphonomy and its relevance to the investigation of large felid involvement in human forensic cases. Forensic Sci. Int. 144 (1), 37−44.

Pokines, J., 2014. Faunal dispersal, reconcentration, and gnawing damage to bone in terrestrial environments. In: Pokines, J., Symes, S. (Eds.), Manual of Forensic Taphonomy. CRC Press, Boca Raton, pp. 201−248.

Reeves, N.M., 2009. Taphonomic effects of vulture scavenging. J. Forensic Sci. 54 (3), 523−528.

Ricketts, D.R., 2013. Scavenging effects and scattering patterns on porcine carcasses in Eastern Massachusetts. Unpublished Masters Thesis. School of Medicine, Boston University, Boston.

Reid, F.A., 2006. The Princeton Field Guide to Mammals of North America. Houghton Mifflin Harcourt, Boston.

Rippley, A., Larison, N.C., Moss, K.E., Kelly, J.D., Bytheway, J.A., 2012. Scavenging behavior of Lynx rufus on human remains during the winter months of southeast Texas. J. Forensic Sci. 57 (3), 699−705.

Smith, J.K., 2015. Raccoon scavenging and the taphonomic effects on early human decomposition and PMI estimation. Unpublished Masters thesis. Department of Anthropology. University of Tennessee, Knoxville.

Sorg, M., Haglund, W., Wren, J., 2012a. Current research in forensic taphonomy. In: Dirkmaat, D. (Ed.), A Companion to Forensic Anthropology. Wiley-Blackwell, Malden, pp. 477−498.

Sorg, M.H., 2011. Scavenging impacts on the progression of decomposition in northern New England. Proc. Am. Acad. Forensic Sci. 17, 384−385.

Sorg, M.H., Haglund, W.D., Wren, J.A., Collar, A., 2012b. Taphonomic impacts of small and medium-sized scavengers in northern New England. Proc. Am. Acad. Forensic Sci. 18, 400−401.

Spradley, M.K., Hamilton, M.D., Giordano, A., 2012. Spatial patterning of vulture scavenged human remains. Forensic Sci. Int. 219, 57−63.

Steadman, D.W., Dautartas, A., Mundorff, A., Vidoli, G., Jantz, L., 2016. Differential raccoon scavenging among pig, rabbit, and human subjects. Proc. Am. Acad. Forensic Sci. 22, 192.

Synstelien, J.A., 2013. Raccoon modification of human skeletal remains. Program of the 82nd Annual Meeting of the American Association of Physical Anthropologists, Knoxville, p. 268.

Synstelien, J.A., Klippel, W.E., 2005. Raccoon (*Procyon lotor*) foraging as a taphonomic agent of soft tissue modification and scene alteration. Proc. Am. Acad. Forensic Sci. 11, 333−334.

Virchow, D., Hogeland, D., 1994. Bobcats. In: Hygnstrom, S.E., Timm, R.M., Larson, G.E. (Eds.), The Handbook: Prevention and Control of Wildlife Damage. University of Nebraska, Lincoln, pp. C35−C43.

Walde, A.D., Delaney, D.K., Harless, M.L., Pater, L.L., 2007. Osteophagy by the desert tortoise (*Gopherus agassizii*). Southw. Nat. 52 (1), 147–149.

Willey, P., Snyder, L., 1989. Canid modification of human remains: implications for time-since-death estimations. J. Forensic Sci. 34 (4), 894–901.

Wright, J.D., Ernst, J., 2004. Wolverine, *Gulo gulo luscus*, resting sites and caching behavior in the boreal forest. Can. Field Nat. 118 (1), 61–64.

Young, A., Márquez-Grant, N., Stillman, R., Smith, M.J., Korstjens, A.H., 2015a. An investigation of red fox (*Vulpes vulpes*) and Eurasian badger (*Meles meles*) scavenging, scattering, and removal of deer remains: forensic implications and applications. J. Forensic Sci. 60 (S1), S39–S55.

Young, A., Stillman, R., Smith, M.J., Korstjens, A.H., 2015b. Scavenger species-typical alteration to bone: using bite mark dimensions to identify scavengers. J. Forensic Sci. 60 (6), 1426–1435.

APPENDICES

These appendices are provided for students, researchers, or investigators with limited background in osteology or ecology. Appendix A is a glossary that includes general terminology, which is organized alphabetically and contains definitions for many of the scientific terms used throughout the text. Appendix B includes definitions of anatomical terminology, as well as a brief overview of the bones of the human skeleton, which can be used to make sense of the descriptions of taphonomic signatures provided in Chapter 5, *What Big Teeth You Have: Taphonomic Signatures of North American Scavengers*. Appendix B concludes with an abridged, annotated bibliography of several osteological and taphonomic texts recommended for those who are interested in learning more.

APPENDIX A: GLOSSARY

Adaptation (n.): specialized characteristic that enables an organism to efficiently perform certain behaviors, such as teeth that can efficiently shear flesh

Adipocere (n.): a greasy, white-colored substance that forms as a byproduct of decomposition from the body's fat, typically in moist environments

Antemortem (adj.): refers to soft tissue wounds or skeletal trauma that occurred before an individual's death, as evidenced by signs of healing

Anthropogenic (adj.): relating to human activity

Appendage (n.): a limb, either arm or leg

Autolysis (n.): enzymatic breakdown of cells that occurs after death, when cellular mechanisms of maintenance and repair have ceased

Behavioral plasticity (n.): the ability to adjust behavior patterns to environmental or social stimuli

Bloat (n.): an early stage of decomposition, evidenced by a marked swelling of the abdominal area, due to gas build-up as a byproduct of the actions of bacteria

Cache (vb.): the act of storing or concealing a food item to prevent spoilage or loss to a competitor, or **(n.)** the location where one stores such a food item

Carnassial teeth (n.): modified fourth upper premolar and lower first molar in carnivores, used to shear and tear flesh

Carrion (n.): dead flesh of a carcass utilized as a food source by scavengers

Competition (n.): the presence of limited resources such as food puts organisms of the same and different species in a struggle with each other, as each individual attempts to gain preferential access to such resources

Conspecifics (n.): organisms of the same species

Crepuscular (adj.): most active at dawn and dusk

Diurnal (adj.): most active during daylight hours

Ecology (n.): branch of biology that focuses on associations between living organisms and their environments, including relationships with other living organisms and influences by nonliving variables (e.g., climate)

Evolution (n.): changes that occur to biological organisms through time due to a variety of factors that can result in the creation of new species after many generations

Facultative scavenger (n.): an organism with a feeding strategy that includes, but does not solely depend on, carrion

Forage (vb.): to seek out wild food resources

Guild (n.): organisms within a community that exploit the same resources

Inertial feeding (n.): manner of feeding used by reptiles and many bird species in which the movement of the head generates momentum to toss food items back into the throat, allowing them to be swallowed

Interspecific (adj.): occurring between *different* species

Intraspecific (adj.): occurring among individuals of the *same* species

Invertebrate (n.): an organism with no internal skeletal system, such as insects and many aquatic organisms including crustaceans and sponges

Kennel pattern (n.): pattern of damage on bones produced by domestic dogs or captive wild animals, where the bone is excessively damaged past the point of being a food source

Kleptoparasitism (n.): a form of competition in which one species or individual usurps a food resource obtained by another

Mesopredator (n.): a medium-sized animal that is both a predator and prey species

Microorganism (n.): an organism that is invisible to the naked eye, including bacteria and archaea

Morphology (n.): study of an organism's physical form

Niche (n.): describes the role an organism plays in its environment including the resources used and how those resources are allocated to survival and reproduction

Nocturnal (adj.): most active at night

Obligate scavenger (n.): an organism with a feeding strategy that is almost exclusively dependent on carrion

Osteology (n.): the scientific study of the anatomy, structure, and function of the skeletal system

Osteophagy (n.): the act of consuming bone, typically to increase dietary intake of minerals such as calcium, sodium, and phosphorous

Perimortem (adj.): refers to wounds or skeletal trauma that occurred around the time of death. In soft tissue, this is evidenced by vital reactions. In bones, characteristics of whether the bone was fresh, or wet, at the time of the insult are evaluated. Bones can retain their organic content making them look fresh for months to even years after death depending on the environment, so while an insult may appear perimortem, it may indeed have occurred during the postmortem period.

Physiology (n.): study of the physical and chemical mechanisms responsible for the normal function of living systems, including whole organisms, organs, tissues, and individual cells

Postmortem (adj.): damage to the body after an individual has died

Postmortem interval (PMI) (n.); also known as time since death (TSD): the time elapsed between death and discovery/analysis

Resource partitioning (n.): division of a niche that reduces competition between organisms within it and promotes their coexistence; for example, bird species which feed on the same insects but live in different levels of the forest canopy, avoiding direct competition

Resource pulse (n.): a significant resource that appears in the ecosystem suddenly and sporadically, but is quickly depleted; carrion is an example

Scatter-hoarder (n.): an animal that has a tendency to both scatter and cache food resources throughout a territory

Taphonomic agent (n.): a living (i.e., animals, plants) or nonliving (i.e., moving water, solar radiation) entity responsible for changes to a body after death

Taphonomic signature (n.): a suite of characteristics that typify a given taphonomic agent

Taphonomy (n.): the study of changes that occur to an organism after its death, including mechanisms of decay, alteration, and dispersal

Taxon (n., *pl. taxa*): a biologically defined group of living organisms such as a species, genus, or family

Taxonomy (n.): the branch of science used to identify, describe, name, and classify biological organisms; also the actual classification scheme of organisms

Trophic level (n.): a group of species in an ecosystem with similar relationships to the primary energy source, and thus having the same position in a food web (e.g., primary producers, herbivores, predators)

Vertebrate (n.): an organism with an internal skeletal system featuring a vertebral column, or backbone, including birds, mammals, reptiles, and fish

Vertebrate scavenging window (n.): the period following death in which carrion is edible to vertebrate scavengers, before the critical accumulation of microbial toxins and the colonization of invertebrates

APPENDIX B: ANATOMICAL AND OSTEOLOGICAL BACKGROUND

Anatomical Directions

Anatomical position (n.): position of the body used when referencing relative locations of anatomical features, pathology, or trauma. Anatomical position in humans consists of the body facing forward, eyes forward, feet parallel with toes forward, and arms slightly out to the sides with the palms facing forward. The terms listed below apply to both humans (bipeds) and quadruped animals unless otherwise indicated.

Anterior (adj.): toward the front of the body (used for bipeds, such as humans). Its opposite term is *posterior*.

Caudal (adj.): toward the tail; used for quadruped animals. Its opposite term is *cranial*.

Cranial (adj.): toward the head; used for quadruped animals. Its opposite term is *caudal*.

Distal (adj.): applies to the upper and lower limbs of the body— means further away from that limb's point of attachment to the torso. For example: the fingers are distal to the elbow. Its opposite term is *proximal*.

Inferior (adj.): toward the sole of the foot (used for bipeds, such as humans). Its opposite term is *superior*.

Lateral (adj.): toward the sides of the body, i.e., further from the body's midline. Its opposite term is *medial*.

Medial (adj.): toward the midline of the body, an imaginary plane that runs along the vertebral column and bisects the body into identical right and left halves (e.g., the breastbone is located in the exact midline). Its opposite term is *lateral*.

Posterior (adj.): toward the back of the body (used for bipeds, such as humans). Its opposite term is *anterior*.

Proximal (adj.): applies to the upper and lower limbs of the body—means closer to that limb's point of attachment to the torso. For example: the shoulder is proximal to the elbow. Its opposite term is *distal*.

Superior (adj.): toward the top of the head (used for bipeds, such as humans). Its opposite term is *inferior*.

Anatomy of a Bone

Cortical bone (n.): the hard, smooth bone that lines the exterior of all bones, and surrounds trabecular bone or the marrow cavity

Diaphysis (n., *pl. diaphyses*): the shaft of a long bone

Epiphysis (n., *pl. epiphyses*): the bulging, irregularly-shaped ends of a long bone; or other features of a bone that develop after the main part forms

Trabecular bone (n.): also known as spongy bone: the internal network of bone, primarily located at the epiphyses of long bones, inside vertebrae and other irregularly shaped bones, and inside some of the bones of the skull

Dental Terminology

Carnassial teeth (n.): the sharp, enlarged upper premolar and lower molar of a carnivore that articulate together when the mouth is closed, adapted to shear flesh; diagnostic for carnivore species

Dental formula (n.): an expression of the number and types of teeth in one quarter of the mouth; in mammals includes the number of incisors (I), canines (C), premolars (P), and molars (M) in one half of the maxilla and one half of the mandible

Heterodont (n.): an organism with multiple types of teeth, each with a distinctive form and function, as in humans or dogs

Homodont (n.): an organism that has one type of tooth, so that all teeth have similar form and function, as in sharks or alligators

Overview of the Human Skeleton

The human skeleton is divided into two major sections: (1) axial and (2) appendicular. The axial skeleton consists of the skull (cranium and mandible, or jaw bone), the vertebral column, the breast bone (sternum), and the rib cage. It also contains the ear ossicles (small bones located in the inner ear) and the hyoid (a bone in the throat). The appendicular skeleton consists of all the other bones: those of the limbs (arms and legs), hands and feet, shoulders, and hips. Table A.1 lists common and scientific bone names with their singular and plural grammatical forms indicated; as well as how many of each bone are typically present in an adult human skeleton.

Table A.1 The Human Skeleton

Skeletal Division	Common Name	Scientific Name	Paired?	# Left	# Right	Total #[a]
Axial skeleton	Skull	Skull or cranium[b] (*pl.:* crania)	No	–	–	1
	Jaw bone	Mandible(s)	No	–	–	1
	Spine or spinal column	Vertebra (*pl.:* vertebrae)	No	–	–	24[c]
	Breast bone[d]	Manubrium (*pl.:* manubria)	No	–	–	1
		Sternum (*pl.:* sterna)	No	–	–	1
	Ribs	Ribs	Yes	12	12	24
	Ear bones[e]	Ear ossicles	Yes	3	3	6
	Neck bone	Hyoid	No	–	–	1
Appendicular skeleton	Shoulder blade	Scapula (*pl.:* scapulae)	Yes	1	1	2
	Collarbone	Clavicle(s)	Yes	1	1	2
	Upper arm bone	Humerus (*pl.:* humeri)	Yes	1	1	2
	Lower arm or Forearm bones	Radius (*pl.:* radii)	Yes	1	1	2
		Ulna (*pl.:* ulnae)	Yes	1	1	2
	Wrist bones[f]	Carpals	Yes	8	8	16
	Hand bones	Metacarpals	Yes	5	5	10
	Finger bones	Hand phalanx (*pl.:* phalanges)	Yes	14	14	28
	Hip bone[g]	Innominate(s) or os coxae (*pl.:* ossa coxae)	Yes	1	1	2
	Lower back bone	Sacrum (*pl.:* sacra)	No	–	–	1
	Tail bone	Coccyx	No	–	–	1
	Thigh bone	Femur (*pl.:* femora)	Yes	1	1	2
	Shin bone	Tibia (*pl.:* tibiae)	Yes	1	1	2
	Lower leg bone	Fibula (*pl.:* fibulae)	Yes	1	1	2
	Kneecap	Patella (*pl.:* patellae)	Yes	1	1	2
	Ankle bones[h]	Tarsals	Yes	7	7	14
	Foot bones	Metatarsals	Yes	5	5	10
	Toe bones	Foot phalanx (*pl.:* phalanges)	Yes	14	14	28

[a]*The total number of the type of bone listed that can be found in a single adult individual.*

[b]*Cranium refers to the skull without the mandible, whereas skull refers to both the cranium and mandible together. Adult crania are made up of multiple fused bones; refer to osteological literature sources for more details.*

[c]*There are three morphologically distinct type of vertebrae: (1) cervical or neck (*n = 7*), (2) thoracic or chest (*n = 12*), and (3) lumbar or lower back (*n = 5*).*

[d]*The two components of the breastbone, the sternum and manubrium, are sometimes fused in adults.*

[e]*The ossicles of the inner ear are typically too small to be of forensic significance.*

[f]*Each wrist has eight carpals: scaphoid(s), lunate(s), triquetral(s), pisiform(s), trapezium (trapezia), trapezoid(s), capitate(s), and hamate(s). Refer to osteological literature sources for more details.*

[g]*Innominates are made up of three fused bones: the ischium (ischia), the ilium (ilia), and the pubis (pubes). Refer to osteological literature sources for more details.*

[h]*Each ankle has seven tarsals: calcaneus (calcanei), talus (tali), navicular(s), cuboid(s), and three cuneiforms. Refer to osteological literature sources for more details.*

FURTHER READING

Christensen, A.M., Passalacqua, N.V., Bartelink, E.J., 2014. Forensic Anthropology: Current Methods and Practice. Academic Press, San Diego, CA.

A comprehensive introductory textbook that covers all aspects of forensic anthropology to include the biological profile (age-at-death, sex, ancestry, and stature) as well as other relevant aspects of the field, such as positive identification and skeletal processing techniques.

DiGangi, E.A., Moore, M.K. (Eds.), 2013. Research Methods in Human Skeletal Biology. Academic Press, San Diego, CA.

In addition to discussing research methods, this volume includes chapters on estimating age-at-death, sex, ancestry, and stature for human skeletons, as well as trauma analysis and taphonomy.

Haglund, W.D., Sorg, M.H. (Eds.), 1997. Forensic Taphonomy: The Postmortem Fate of Human Remains. CRC Press, Boca Raton, FL.

Essential reading as the first text to be written on applications of taphonomy to forensic contexts with material that remains relevant.

Haglund, W.D., Sorg, M.H. (Eds.), 2002. Advances in Forensic Taphonomy: Method, Theory, and Archaeological Perspectives. CRC Press, Boca Raton, FL.

The follow-up text to the above Haglund and Sorg (1997) volume.

Langley, N.R., Tersigni-Tarrant, M.T. (Eds.), 2017. Forensic Anthropology: A Comprehensive Introduction. CRC Press, Boca Raton, FL.

Introductory textbook that details biological profile analyses (age-at-death, sex, ancestry, and stature) in addition to pertinent chapters on forensic taphonomy and forensic archaeology.

Pokines, J.T., Symes, S.A. (Eds.), 2014. Manual of Forensic Taphonomy. CRC Press, Boca Raton, FL.

This comprehensive text contains a variety of information on different taphonomic agents, environmental effects, and timing of trauma in a taphonomic context.

Schotsmans, E.M., Márquez-Grant, N., Forbes, S. (Eds.), 2017. Taphonomy of Human Remains: Forensic Analysis of the Dead and the Depositional Environment. Wiley-Blackwell, Hoboken, NJ.

An extensive and comprehensive volume, covering taphonomic processes and agents essentially from A to Z, from entomology to mass graves to scavengers to time since death to case studies, and everything in between.

White, T.D., Folkens, P.A., 2005. The Human Bone Manual. Academic Press, San Diego, CA.

An abridged version of White et al. (2012) below—good for field contexts and nonexperts.

White, T.D., Black, M.T., Folkens, P.A., 2012. Human Osteology, third ed. Academic Press, San Diego, CA.

One of the most comprehensive texts for adult human skeletal anatomy, useful for biological anthropologists, anatomists, forensic scientists, and others interested in human skeletal morphology.

INDEX